# A QUOTIDIAN QUASH
## From Mental Hygiene to Mental Health
## 1969–2012

Part One, Part Two, and Part Three

DORIAN REDUS

Copyright © 2022 Dorian Redus
All rights reserved
First Edition

PAGE PUBLISHING
Conneaut Lake, PA

First © Copyright 2013 Dorian Gaylord Redus.

All rights reserved. No part of this publication may be reproduced, stored in a retrieval system, or transmitted, in any form or by any means, electronic, mechanical, photocopying, recording, or otherwise, without the written prior permission of the author.

Library of Congress Control Number: 2013902458

ISBN 978-1-6624-7578-8 (pbk)
ISBN 979-8-88793-662-8 (hc)
ISBN 978-1-6624-7579-5 (digital)

Printed in the United States of America

In memory of my mother, Vivian Fay Stinette-Redus, and my editor(s).

People like their
stories are uniquely
beautiful because
of the *details*.

# PREFACE

Any concerned humanitarian may find this book a useful book to read. According to this book, as surely as you receive all the very carnal therapy in the information relationships via words in a "relativistic color television universe," the adversarial and concerned may see or find care or harm, fairness or cheating, liberty or oppression, loyalty or betrayal, authority or subversion, sanctity or degradation, and rape care with harm, fairness with cheating, liberty with oppression, loyalty with betrayal, etc. until as in this book, the adversarial discipline of psychiatry has amassed insanity and crimes against me greater than my insanity and crimes against my society, which psychiatry is at preying on because psychiatry can't tell "right from wrong," and my judges can't find "guilt or innocence" without my cosmological TV *energy* being presented via this book. The six pairs of moral concerns above are from Jonathan Haidt's *The Righteous Mind* (2012).

Furthermore, in accordance with a most necessary decree (from above that protects the real psychiatric agents and the real psychiatric agencies that are both powerful and above the law), most of the real nonfiction letters in this book use fictitious names that protect the guilty. Conversely, in this book, the many ideas in its many letters to California's guilty therapists still manage to add to a therapeutic and (urgent) stream of the author's conscious psychiatric and inpatient voice. Finally, this book, word by word, does not hide the author's very defensive needs. Enjoy!

# A QUOTIDIAN QUASH: FROM MENTAL HYGIENE TO MENTAL HEALTH

Mr. Dorian Gaylord Redus
Ward T-15
Napa State Hospital
2100 Napa-Vallejo Hwy.
Napa, CA 94558-6234
1(707)252-9988

Wednesday, February 29, 2012

The Honorable Master Calendar Judge
San Francisco County Superior Court

Re: the fixed maintenance of my sanity.

Court Number SC088778     CII: M02858702
Maximum Commitment Date: Until my sanity is restored

To Whom It May Concern:

For decades, going on forty years, I have agreed to disagree with San Francisco's paragon courts. However, with each new and coming year, I assert my case for my sanity more cogently and more formidably. This whole word-processed document, in its entirety, is my statement to my court officers for 2012. Love is giving what you want to give; happiness is wanting what you get. My courts have made me happy for forty years. This year, with this court document, I give what I want to give: my vehicle of sanity. I hope to win my legal sanity as a consequence of all my court officers reading this document.

In this binding, there is the consequential six-page November 15, 2010, letter to my (public defender) attorney-at-law, Cheryl H. Arkansas. And in this binding, there is the stellar twenty-page treatise, *A Three-Part Discussion Including Part Four*, written on Tuesday, January 17, 2012. Moreover, there are the authoritative January 7, 2011, letter about alleged child molesting and the mandatory letter that I wrote to the late Reverend Dr. Martin Luther King Jr., after his untimely demise such that I draw the attention to the letter's issue in hopes of an appropriate reply or response, which I never got.

Almost thirty-five years after my PC 1026, for my 1974, PC 187, it is not as if anyone was unsafe, actually physically hurt, or harmed by the vicious voices that caused me to return to Napa on October 1, 2009. Furthermore, the vicious voices have stopped, and so far they have not returned. They just stopped just before my return to Napa. Perhaps they just stopped due to my psychotropic medication (Risperidone) being increased, at my request, just before my return. *But a long vacation from my outpatient treatment program was, it seems, necessary for me to heal, regain my sanity, and write this long document. So being here or somewhere writing may have been the real and necessary healing trick.* It was not physically possible to write as well as I have here until I was in an open ward. However, I have continually been at Napa State from October 1, 2009, to present. Thank you.

Respectfully submitted,

Mr. Dorian Gaylord Redus
Psychiatric patient

# A QUOTIDIAN QUASH: FROM MENTAL HYGIENE TO MENTAL HEALTH

Mr. Dorian G. Redus
Ward T-15
Napa State Hospital
2100 Napa Vallejo Hwy.
Napa, CA 94558
707.252.9988

Wednesday, September 29, 2010

Cory Alabama, CSW
Ward T-15
Social Worker
Napa State Hospital
2100 Napa Vallejo Hwy.
Napa, CA 94558

Dear Social Worker:

Someone in my dorm turned the light on at four in the morning, and this is a brief historic letter of my ineluctable stream of consciousness around 4:15 a.m., Tuesday, September 21, 2010. I have never molested a child, not even my infant nephew, Jason Biller, in 1967, my last reported allegation. I often said I molested for over three years. I was in my San Francisco sex offender's group, which ended in early 2009, but I was just being cooperative and just complying. Again, I have never molested. I was just being curious about why molesters molest children. I am also curious about why Ed Utah molested for so long and yet recently directed this hospital as its executive director. I am curious why Dr. Vernon Indiana, MD, my first Napa State Hospital psychiatric examiner, committed suicide in the 1970s. I am curious why the late Dr. Donald Montana, MD, my first VA psychiatrist, told me to get a gun before my 1974 murder. And I am curious why he never appeared (for or against) me in a California Court of law.

I am curious why there has been so much confabulatory "free fabrication" in the testimony of my prosecutors and from their court officers. And furthermore, I am curious as to why there is so much San Francisco Conditional Release Program discretion in my forensic affairs.

I have not had heterosexual intercourse (sex) since I had sex with my then wife, now ex-wife in the 1990s. And I have not had any form of homosexual sex since the 1980s. I may be curious and queer-looking, but I am not a homosexual. Although, due to the medication Prolixin's torturing side effects, I had sex with about one hundred men in about a year, in about 1981 on Ward 27, at Atascadero State Hospital. And due to the medication Haldol's torturing side effects, I had sex at Napa State Hospital with about one hundred men on Ward Q3 and 4, in about a year in about 1984.

I hope this helps us get my therapy going (or better). I hope that this letter is at least an incipient conversation between us.

In conclusion, I am curious as to why I killed my instant offense victim, Ms. Edna Ella Robenson, on August 9, 1974. We all know my side. I have said it was self-defense, iatrogenic (doctor caused) delusions, and paranoia. However, I now need "the whole truth and nothing but the truth," and my attorney-at-law and my California Superior Court seem to be unable to help!

Thank you,

Mr. Dorian G. Redus

Cc: Cheryl H. Arkansas
    Attorney-at-Law
    214 Duboce Avenue
    San Francisco, CA 94103
    415.431.0425
Fax 415.255.8631

PS I am no longer hearing the angry (alliterative) vicious voices that caused me to return to this state hospital on October 1, 2009.

# A QUOTIDIAN QUASH: FROM MENTAL HYGIENE TO MENTAL HEALTH

Mr. Dorian G. Redus
Ward T-15
Napa State Hospital
2100 Napa Vallejo Hwy.
Napa, CA 94558
707.252.9988

Wednesday, October 20, 2010

Dr. Florida, MD
Ward T-15
Psychiatrist
Napa State Hospital
2100 Napa Vallejo Hwy.
Napa, CA 94558

Dear Doctor:

This complaint is that my Golden Gate Conditional Release Program's April 12, 2010, annual report and revocation of my outpatient treatment letter to the Honorable Judge Wyoming was uncontested, and thus my last court hearing was, as my attorney put it to me, "at the discretion of CONREP." without me being there to state the following—and other complaints about my Conditional Release Program's letter of April 12, 2010.

In their April 12, 2010, letter to my court, Dr. May, PhD, and Christopher A. Idaho, LCSW, have such a loose relationship with the truth and the fact of their letter, and thus their input has been a travesty of justice. In all fairness to all of us, I must say that they copied many of their errors from the previous Conditional Release Program director's errors. Dr. Porky, PhD, also had a loose use of the truth and the fact. Furthermore, my community program director, Christopher A. Idaho, and my clinician, Dr. May's, daily and weekly treatment put words in my mouth. For example, I reported to the authorities a score of years ago that Dr. Donald Montana, MD, the San Francisco Veterans Administration Chief of Mental Hygiene, in the 1970s, told me to "Get a gun!" Perhaps he wanted to fight me. I have never known, and he was never (I guess he was above the law) brought to my court to testify, like I needed, on the issues. My complaint is that my CONREP(s) regularly misquoted me as saying that I said Dr. Montana said to get a gun and shoot Ms. Edna Ella Robenson, my instant offense victim. That misquote puts the shoes of the wrongdoer and mental illness on me in an ironic and treacherous travesty of justice. Furthermore, in continuing exacerbation of my forensic and psychiatric affairs and condition respectively, social worker Chris Idaho of San Francisco CONREP had me saying the whole thing was "originally" my delusion. He put words in my mouth, leading me to say I was delusional over and over in Teams.

My two San Francisco CONREPs have been too deleterious and too full of confabulation like I have elucidated above. My San Francisco court hearings are pejorative, and they are very complicated. I need your help! When I stand up for myself, my CONREP(s) say I am decompensating!

In conclusion, this issue is just one of about fifteen complaints I have regarding the sometimes misleading April 12 letter. I must somehow stop the loose relationship to the truth and the loose relationship to the fact.

Sincerely yours,

Mr. Dorian G. Redus

Cc: Cheryl H. Arkansas
    Attorney-at-Law
    214 Duboce Avenue
    San Francisco, CA 94103
    415.431.0425
Fax 415.255.8631

Mr. Dorian G. Redus
Ward T-15
Napa State Hospital
2100 Napa Vallejo Hwy.
Napa, CA 94558
707.252.9988

                                             Monday, October 25, 2010

Dr. Hameed Nebraska, MD
Ward T-14
Psychiatrist
Napa State Hospital
2100 Napa Vallejo Hwy.
Napa, CA 94558

Dear Doctor:

    Thank you for speaking to me about a medication holiday on Friday, October 22, 2010, as we serendipitously passed by each other in front of Ward T-15.
    Kind sir, may I please send to you three current letters that I recently wrote here on Ward T-15? They are letters that any doctor involved in helping me with my forensic or psychiatric decisions should have to read first.

Sincerely yours,

Mr. Dorian G. Redus

Mr. Dorian G. Redus
Ward T-15
Napa State Hospital
2100 Napa Vallejo Hwy.
Napa, CA 94558
707.252.9988

Saturday, October 30, 2010

Dr. Mexico, PhD
Ward T-15
Psychologist
Napa State Hospital
2100 Napa Vallejo Hwy.
Napa, CA 94558

Dear Doctor:

Please! Who should come first? How should we all prioritize and groom me, the "once" homicidal mental patient? Not who came first, the chicken or the egg—a chronological thing. The VA came first after the US Army, and others have since the US Army manipulated my courts as if they are all kangaroo courts. Doctors, "others who are systemic and paid to," have manipulated my court hearings, my mental hygiene programs, and my San Francisco Community Outpatient Treatment Programs of the past. In the future, I do not want to reoffend, but I need to take the wheel and to drive us all to our goal—a mentally healthy Mr. Dorian G. Redus. My ways to our goals need to be first on our list of our priorities.

First of all, "Go San Francisco Giants" to win the 2010 World Series events! I know myself well. I am intrinsically an outstanding color television spectator, and when my courts, my San Francisco Conditional Release Programs, and my hospital, Napa State Hospital, all respect me and give me the respect of a court (disabled) dignitary, I will also prove to be outstanding, like my Admission Ward Psych testing has already shown. My results were *High Average—Superior range of intelligence* (on 10/07/09, Dr. Dakota, p. 4 of 29, Ward T-2, 11/24/2009, Wellness and Recovery Plan).

Before, when I preferred my own druthers, I was coerced to try the San Francisco Golden Gate Conditional Release Program's community outpatient treatment program by Dr. Douglas Porky. I was coerced to try the San Francisco Golden Gate Conditional Release Program's community outpatient treatment program by Christopher Idaho, LCSW. And I now behest, ensconced on this ward, Ward T-15 of Napa State Hospital in California, that I try my San Francisco Golden Gate Conditional Release Program's community outpatient treatment program with my priorities. At sixty-four years of age with an adult daughter and two grown granddaughters, it is three strikes, and you are out in California!

I put in a written general request on Friday, October 29, 2010, hoping to keep my right to mail and to send my written *Redus Treatise* (it is an essay of sorts) to my last university, San Francisco State University, where I am a junior. Will you please help me in this endeavor to communicate by giving it your personal approval and your personal approbation?

Please? Do it as an open ward privilege—post hoc, ergo propter hoc—after my maintaining of my G-Card and because of my maintaining of my G-Card.

Sincerely yours,

Mr. Dorian G. Redus

Cc: Cory Alabama, CSW, Ward T-15

Mr. Dorian G. Redus
Ward T-15
Napa State Hospital
2100 Napa Vallejo Hwy.
Napa, CA 94558
707.252.9988

                                  Halloween, Sunday, October 31, 2010

Dr. Florida, MD
Ward T-15
Psychiatrist
Napa State Hospital
2100 Napa Vallejo Hwy.
Napa, CA 94558

Dear Doctor:

    This complaint is that my San Francisco Golden Gate Conditional Release Program's April 12, 2010, annual report and revocation of my outpatient treatment letter to the Honorable Judge Wyoming was uncontested, and thus my last court hearing was, as my attorney put it to me, "at the discretion of CONREP" without me being there to state the following—and other complaints about my Conditional Release Program's letter of April 12, 2010.
    In their April 12, 2010, letter to my court, Dr. May, PhD, and Christopher A. Idaho, LCSW, have such a loose relationship with the truth and the fact of their letter, and thus their input has been a big travesty of justice. In all fairness to all of us, I must say that they copied many of their errors from the previous Conditional Release Program director's errors. Dr. Porky, PhD, also had a loose use of the truth and the fact. I have a specific bone of contention and, in general, a bone to pick. Currently, my specific subject of dispute and my general grounds for complaint are both found, like watered seeds growing, in the following quote from pages 4 and 5 of my former San Francisco outpatient treat-

ment program's letter of April 12, 2010, requesting my outpatient treatment be revoked:

> During an individual session with Mr. Redus (dated 6/9/09), he was fixated on a paper he wrote in the 1970's for an Astronomy class. He went into great details about finally finding proof to prove his theory that space is unfolding. He stated that he has been working on this paper off and on for the past 30 years. He was upset that his instructor at the time refused to grade his paper for extra credit and filed a complaint with the department… Hence, it is important to continue to monitor his delusional thoughts as it may lead to anger. (Pages 4 and 5 of the April 12, 2010, letter)

I do have a theory that we here on earth are part of a big cosmic RCTVU (relativistic color television universe), and I do have a theory that we here on earth are part cosmic STS (space-time sphere). Moreover, I find working on these two theories invigorates my superior powers of mind and intelligence. However, I also find my former San Francisco outpatient treatment program's intelligence, information, and news on my work amounts to a subterfuge, a "deceptive stratagem," and I take umbrage at their intelligence. And furthermore, I take even greater umbrage, offense, at their tutelage because they do not see that I need a university tutor and perhaps some brainstorming networking professors helping me with my theories for our mutual improvement and support. I need for my theories (RCTVU and STS) to be like watered seeds growing. The next appropriate step is to send my treatise on my two theories to my university, San Francisco State University, for an academic evaluation.

*First* of all, when I took classes from him in the early 1970s, City College of San Francisco's Professor Edwin Duckworth was an extraordinary instructor with support from world-class brainstorming and networking extracurricular (scientific) summer seminars in

1972, 1973, and 1974 that were televised on KQED's Channel 9. I learned a lot from him and from his San Francisco Palace of Fine Arts weekly visiting lecturers from around the nation. They were the best. One day in astronomy class, Mr. Duckworth said, "I will grade an extra credit paper on anything." Perhaps I stumped him. I do not know.

I wrote a short paper on Dr. Edwin Hubble's Law and Dr. Einstein's special relativity. In my brief work, I wrote that according to Hubble's Law, higher and higher distances separating galaxies and clusters of galaxies produce greater and greater recession velocities until—it seems, according to special relativity—you are limited by the speed or velocity of light. The professor called it the *Redus Treatise*, and he refused to grade it. I believe it was 1972. Mostly, I remember City College of San Francisco's 1972 summer seminar directed by my professor, Mr. Edwin L. Duckworth, *Stellar Evolution: Man's Descent from the Stars*, and I also remember that I was twenty-six years old.

*Second,* the following is from a book by Dr. Mario Livio, PhD, and he is with the Hubble Space Telescope Program. Eureka! The find follows:

> Second, I would like to clarify that it is only the scale of the universe at large, as expressed by the distances that separate galaxies and clusters of galaxies, that is expanding. The galaxies themselves are not increasing in size, and neither are the solar systems, individual stars, or humans; space is simply *unfolding* between them. Furthermore, a common misunderstanding is to think of the galaxies as if they are *moving* through some pre-existing space. This is not the case. Think of the dots on the surface of the balloon. Those dots are not moving at all on the surface (which is the only *space* that exists). Rather, *space itself* (the surface) is stretching, thus increasing the distances between galaxies. Finally on this point, the limit for special relativity that matter cannot move

faster than light does not apply to the speeds at which galaxies are separated from each other by the stretching of space. As I explained above, the galaxies are not really moving, and there is no limit on the speed with which space can expand.[1]

*Three*, I do not feel or think that I have a self-love to a fault. I feel and I think that I have a niggling enamor of truth, fact, and above all, a love of knowledge. Focusing once again on the previously mentioned paragraph from pages 4 and 5 of my outpatient treatment program's April 12, 2010, letter, I cut out the following two sentences, put in an ellipsis, and pasted them here as they are a refuted issue:

> When explore [ing] further, Mr. Redus admits that his interest in the topic was a function of his grandiose delusions. He felt a narcissistic injury when his instructor rejected his paper. (Page 5, April 12, 2010, letter)

*Four*, Dr. Wendy May, PhD, omitted my main message, which is that I agree with Dr. Mario Livio. When describing the expansion of the universe from the big bang, one should see and discover that paradoxically, the galaxies are "unmoving." Moreover, when the word *unfolding* is Dr. Livio's observation, term, and theory, the April 12, 2010, letter adduces and points out, as if I was being an idiot, that I have a theory that space is unfolding between galaxies, causing the expansion of the universe.

My two San Francisco CONREPs have been too deleterious and too full of confabulation like I have elucidated above. My San Francisco court hearings are pejorative, and they are very complicated. I need your help! When I stand up for myself, my CONREP(s) say I am decompensating!

---

[1]. Mario Livio, *The Accelerating Universe: Infinite Expansion, the Cosmological Constant, and the Beauty of the Cosmos* (New York: John Wiley & Sons, Inc., 2000), 49

In conclusion, this issue is just one of over twenty complaints that I have regarding the sometimes misleading April 12, 2010, letter. I must somehow stop the unscientific loose relationship to the truth and the loose relationship to the fact.

Sincerely yours,

Mr. Dorian G. Redus

Cc: Cheryl H. Arkansas
    Attorney-at-Law
    214 Duboce Avenue
    San Francisco, CA 94103
    415.431.0425
Fax 415.255.8631

Cc: Cory Alabama, SCW

Mr. Dorian G. Redus
Ward T-15
Napa State Hospital
2100 Napa Vallejo Hwy.
Napa, CA 94558
707.252.9988

                                               Monday, November 15, 2010

Cheryl H. Arkansas
Attorney-at-Law
214 Duboce Avenue
San Francisco, CA 94103
415.431.0425
415.255.8631 Fax

Re: pecuniary remuneration (funding), defendant nonfeasance, and more complaints, discrepancies, in the Monday, April 12, 2010, letter that recommended that my outpatient treatment is revoked. This is a David and Goliath story of the pusillanimous and the puissant.

Dear Attorney:

    This complaint is that my San Francisco Golden Gate Conditional Release Program's Monday, April 12, 2010, annual report and revocation of my outpatient treatment program letter to the Honorable Judge Wyoming was uncontested, and thus my last court hearing was, as you put it to me, "at the discretion of CONREP" without me being there to state the following—and other complaints about my former Conditional Release Program letter of April 12, 2010.
    In their April 12, 2010, letter to my court, Dr. May, PhD, and Christopher A. Idaho, LCSW, have such a loose relationship with the truth and the fact of their letter, and thus their input has been a big travesty of justice. In all fairness to all of us, I must say that they copied many of their errors from the previous Conditional Release

Program director's errors. Dr. Porky, PhD, also had a loose use of the truth and the fact. I have a specific bone of contention and, in general, a bone to pick. Currently, my specific subject of dispute and my general grounds for complaint are both found, like watered seeds growing, in the following two quotes from my former San Francisco outpatient treatment program's letter of Monday, April 12, 2010, requesting my outpatient treatment be revoked:

*The first quote is from page one.*

He [Dorian Redus] had informed his psychiatrist at the time that his girlfriend was having an affair. He heard his psychiatrist respond, "Get a gun and shoot her."

*The second quote is from page four.*

On several occasions, he thought he heard his psychiatrist [Dr. Donald Montana] telling him he should buy a gun to kill her and that his psychiatrist was coaching him to commit murder.

The previous two small quotes above are more grievous and serious than we can suppose. An individual treatment of each, the first and the second quote, follows:

*The first quote is from page one.*

He [Dorian Redus] had informed his psychiatrist at the time that his girlfriend was having an affair. He heard his psychiatrist respond, "Get a gun and shoot her."

The outrageous quote above is a favorite tactic of my San Francisco Conditional Outpatient Treatment Program, and it squelches, squashes, and silences me in court and in my daily life, but it does not squelch me in this letter. If I have said it once, I have said it one thousand times. And saying it one time is too many because, it seems, saying the above is a misleading quote. One thousand times has not been enough to stop the misleading, misquoting, and slander on San Francisco's Superior Court documents or been enough to stop the slander rampant in testimony. Furthermore, this first quote is also keeping me from "laughing all the way to the bank" because of my defense's defendant nonfeasance on the issue of psychiatric malpractice by the Veterans Affairs doctor who told me to "Get a gun."

When, if not now? Has not my defense's defendant nonfeasance on the issue of psychiatric malpractice been from my last hearing back for around two score of years, and is it not continuing into the future of psychiatric malpractice in California—unfair and just plan pejoratively?

*The second quote is from page four.*

> On several occasions, he thought he heard his psychiatrist [Dr. Donald Montana] telling him he should buy a gun to kill her and that his psychiatrist was coaching him to commit murder.

Perjure is sometimes a serious crime. In court, on my own, or telling the whole truth or fact about myself, I have never said, in the past two score of years, that the doctor, Dr. Donald Montana, MD, said anything like the San Francisco Conditional Release Program's statement: "On several occasions, he [Dorian G. Redus] thought he heard his psychiatrist [Dr. Donald Montana] telling him he should buy a gun to kill her [Edna Ella Robenson] and that his psychiatrist was coaching him to commit murder." I did say, two score of years ago and many times since that once or twice, Dr. Donald Montana, MD, told me to, "Get a gun." If I have said it once, I have said it one thousand times, and saying it one time is too many because, it seems, saying the above and previous are misleading quotes. One thousand times has not been enough to stop the misleading, misquoting, and slander on San Francisco Superior Court documents or been enough to stop the unfair slander and ramifying rampant in my courtroom's sworn testimony. Furthermore, both quotes are keeping me from successfully "laughing all the way to the bank" because of my defense's defendant nonfeasance on the issue of courtroom and psychiatric malpractice.

When, if not now? Has not my defense's defendant nonfeasance on the issue of psychiatric malpractice been from my last hearing back for around two score of years, and it is not continuing into the future of psychiatric malpractice in California—unfair and just plan pejoratively.

If the court pays me enough for helping my courts see the whole truth, then the twisting of the truth to make traps for mean people to catch me will stop! I am the originator! Why can't I change what court officers say I said back to what I really said? Isn't all this gross, that my "casting of pearls before swine" has been falling on deaf ears is surely a gross understatement?

*A specific general diatribe on the Monday,*
*April 12, 2010, letter follows:*

(1) The San Francisco California Golden Gate Conditional Release Program community outpatient treatment clinicians, Dr. Wendy May, PhD, and Christopher Idaho, LCSW, got it wrong again in their Monday, April 12, 2010, letter. My crime was on August 9, 1974, the day Nixon resigned the presidency of the United States. They said, incorrectly, that it was August 7, 1974, on page 1 of the April 12, 2010, letter.

*Antecedents to the crime*

(2) The San Francisco California Golden Gate Conditional Release Program community outpatient treatment clinicians, Dr. Wendy May, PhD, and Christopher Idaho, LCSW, got it wrong again in their Monday, April 12, 2010, letter. They used *common-law wife* as a sufficient and an appropriate term for my instant offense victim, Edna Ella Robenson. *Common-law wife* does not describe

our relationship. She gave me many warning signs. Although it was beautiful day, and I was out flirting, my apercu, first thought, my response at seeing Edna was "not with her," and she was well dressed and good-looking. I also—and this is very rare for me—hallucinated that her hair had red stage blood in it. These things were a definite turnoff. These first impressions were bad, but later, things got much worse. Within the first few months, Edna became suicidal once and also attacked me with our kitchen knife, which happened many more times during the six years we were in our doctor-advised relationship. Sometimes she attacked me with our kitchen knife, and then she also called San Francisco Police to our apartment. After all of that for years, I hit her one night, and soon thereafter, when she said the police were getting her a gun, I took her life on the worst morning of my life. Because my Veterans Affairs doctor, Dr. Donald Montana, MD, advised me to stay in the relationship many times, I continued in the relationship over many years of her violence, and there were times I left her and went back because he advised me to go back.

My *jeopardy* during Edna's violent years caused me to think people don't really die—a growing delusion I had throughout her violent years. The delusion stopped with Edna's demise. Edna had been my doctor's ideal common-law wife for me, his patient for six years. Because he lived and died, and he never came to the courtroom where I was, my courtrooms are still in jeopardy of a miscarriage of justice.

Along with most of the above making it difficult to simply relate to Edna as my common-law wife during the whole six years that I was with Edna, I knew a struggling young mother in a city nearby San Francisco with three young ones—one of them mine.

Furthermore, not long before I took Edna's life, murdering her due to "my illness" on August 9, 1974, I went to then Sheriff Richard Kentucky's home around the corner from my parent's San Francisco home, and I asked him for some much-needed help with Edna and my Veterans Affairs doctor. I told the San Francisco Sheriff that Edna was assaultive, and therefore she was in danger as I was getting more and more likely to murder her.

I also gave the sheriff Dr. Donald Montana's name as my psychiatrist. Also, before my instant offense and after the doctor told me to "Get a gun," post hoc, ergo propter hoc, I also tried a Veterans Affairs Hospital near Stanford University, telling my VA doctor there everything. I spilled all the beans. On these matters above on the term my *common-law wife*, I am selectively viewed and assumed to be mistaken, a teller of lies, or a deleterious delusional nut! I am actually just a humiliated and humble sinner—"neither cast ye your pearls before swine."

(3) The San Francisco California Golden Gate Conditional Release Program community outpatient treatment clinicians Dr. Wendy May, PhD, and Christopher Idaho, LCSW, got it very wrong again in their Monday, April 12, 2010, letter. They said on page 1, "Following an argument on the day of his instant offense, he had gone home, changed into overalls, and placed a butcher knife in his pocket." They imply first an argument then a trip home to change, place a weapon, and then the evil murder. This is quite simply not the way it happened. Moreover, the murder was a complicated culmination as I explained clearly above in diatribe number two. What they say happened might have happened before a murder. However, it did not happen their way as they describe it before my August 9, 1974, murder. Their written recount is a mix-up of my real truth and fact—until their description establishes a misleading lie—in print.

(4) The San Francisco California Golden Gate Conditional Release Program community outpatient treatment clinicians Dr. Wendy May, PhD, and Christopher Idaho, LCSW, got it wrong again in their Monday, April 12, 2010, letter. On page 1, they said, "He [Dorian] was consumed with rage and jealousy [at the time of Dorian's instant offense]." Maybe Dr. Montana and Edna Ella Robenson were consumed with rage and jealousy around the time of my instant offense. He said, "Get a gun." As for me, I just thought, quite simply, actually, as I elucidated above in diatribe number two, that things were culminating, and I was fighting

and murdering an aggressor in sinuous self-defense. My jeopardy had reproduced a challenging delusion about death. I mentioned this in diatribe number two that had first emerged in the Army.

(5) The San Francisco California Golden Gate Conditional Release Program community outpatient treatment clinicians, Dr. Wendy May, PhD, and Christopher Idaho, LCSW, got it wrong again in their Monday, April 12, 2010, letter. On the lower part of page 1, they said, "He [Dorian] went to meet with his psychiatrist [the Veterans Affairs doctor] and returned to the scene of the crime later that afternoon." What that does not say is that I had an appointment, and it moreover does not say that during that appointment, I confessed to that morning's crime—no confabulation. And after my confession, the doctor let me walk out of his office on my own, like I was lying to him about having just murdered.

(6) The San Francisco California Golden Gate Conditional Release Program community outpatient treatment clinicians, Dr. Wendy May, PhD, and Christopher Idaho, LCSW, are quite correct in their Monday, April 12, 2010, letter. At the bottom of page 1, they said, "He [Dorian] drove around with her body in a rented pickup truck and contacted his attorney, [Hyrim E. Smith] who arranged for his surrender to police." This is so close to true. Maybe I should not have a complaint. But I drove in mortal fear one night just after the instant offense, contrived a lie to show Edna's demise was due to a regrettable accident that required me to move the corpse, and put it in the pickup truck, which had its bed filled up with a bale of yellow straw. My lie was to burn the truck and the corpse, but I just did not need to add another lie to these august lies by others. So after having confessed to my psychiatrist, I confessed again to my attorney at the law offices of White Badman and Smith of San Francisco. Which is where I get my indelible predilection, preference, for my underpinnings of pathos—pity, sympathy, tenderness, and sorrow caused regarding my August 9, 1974, crime.

(7) The San Francisco California Golden Gate Conditional Release Program community outpatient treatment clinicians Dr. Wendy May, PhD, and Christopher Idaho, LCSW, got it wrong again in their Monday, April 12, 2010, letter. In the middle of page 2, they said:

> Previous reports showed that at the end of 2004, the client requested a session with his previous clinician to confess an incident from when he was eighteen or nineteen years old during which he took his infant nephew into a closet to simulate oral copulation. Mr. Redus said he felt his lips "bump" what must have been the infant's genitalia. Although the incident is far in the past, a CPS report was filed and Mr. Redus completed Sex Offender treatment. He is not a 290 Registrant.

I requested a session with my previous clinician, Dr. Steve Kansas, PhD, to confess an incident from when I was eighteen or nineteen years old, not "out of the blue" but because the day before, there had been a meeting with myself, Dr. Steve Kansas, and the director of my community outpatient treatment program, Christopher Idaho. At that meeting, they said, "We are calling your sister to ask her about child molest." I seriously doubt I touched genitalia. Although, under the scrutiny of Dr. Wendy May, PhD, thinking aloud and mulling over old 1967 memories, I used the word *genitalia* describing the 1967 incident. I have never molested a child, not even my infant nephew in 1967, my last reported allegation. I often said I molested for the over three years I was in my San Francisco Sex Offender's group, which ended in early 2009, but I was just being cooperative and just complying. Again, I have never molested a child or an infant.

(8) The San Francisco California Golden Gate Conditional Release Program community outpatient treatment program clinicians, Dr. Wendy May, PhD, and Christopher Idaho, LCSW, got it

wrong again in the Monday, April 12, 2010, letter. In the middle of page 4, they wrote:

> He has two windows in his bedroom, where one has a screen and the other does not. He makes sure that he does not change his clothes in front of the window without a screen for fear that he may fall out accidentally.

All because my best lady friend at CONREP San Francisco had recently jumped from a window of her Richmond district home and nearly died. She had been coming over religiously, every Sunday, for months for dinner and a movie.

My two San Francisco CONREPs. have been too deleterious and too full of confabulation like I have elucidated above. My San Francisco court hearings are pejorative, and they are very complicated. I need your help! When I stand up for myself, my CONREP(s) say I am decompensating!

In late September and early October 2009, the San Francisco Veterans Affairs psychiatrist Dr. Fred Ohio, MD, told me that I did not need to be hospitalized because I was hearing "voices." The San Francisco community outpatient treatment psychiatrist, Dr. Anthony Texas, MD, also told me, "I don't think you need to be returned to Napa State hospital because you are hearing angry voices. Dorian, you know better than to act on what the voices say. You are not going to act out, so you don't need to go back."

I think San Francisco's current community outpatient treatment program was wrong to have me hospitalized on October 1, 2009. I feel San Francisco's current community outpatient treatment program was wrong to revoke me. I also feel San Francisco's current community outpatient treatment program got it wrong because it believes in Christopher Idaho's absolutism, and the director, Christopher, wanted me to be re-hospitalized at Napa State Hospital on October 1, 2009.

In conclusion, as I have elucidated above, the Monday, April 12, 2010, letter is sometimes misleading. We must somehow stop the unscientific loose relationship to the truth and the loose relationship to the fact. A word to the wise and the sapient regarding the full self-disclosure I am required to make to my Conditional Release Program and to my community outpatient treatment program clinicians. Perhaps this is a referendum of sorts, and perhaps it is an ultimatum of sorts to a tyrant regime of procrustean prosecutors showing little or no regard for my individual differences like my high average-superior range of intelligence or my special circumstances like my average funding.

*Four issues that have made me angry in the recent past*

*One*, my recently retitled scientific paper, *A Three-Part Discussion*, being called my delusion and its being summarily dismissed, as if I am a deleterious delusional narcissistic nutcase. Both of these have made me angry and exacerbated for decades!

*Two*, when I had already and immediately provided the truth, the whole truth, and nothing but the truth, so help me God, my recent three plus years of sex-offender group and treatment made me feel like the authorities felt I was telling them lies. The last allegation was over forty years old.

*Three*, the Veterans Affairs doctor, their Chief of Mental Hygiene in the 1970s, Dr. Donald Montana, MD, and therefore the Veterans Affairs itself getting away with more and more malpractice for two score of years—forty years and counting—is vicious and tending to worsen.

And *four*, the sundry, various, and miscellaneous travesties of justice that are Napa State Hospital's and my CONREP's testimonies in superior courts being lies, nontruths, and nonfactual for the past ten or more years is very vexatious, perplexing, and a real bother.

My (supposed) allegation is that all four of the above culminated in 2009, causing me to hear "angry voices." All four also keep me from simply working with my San Francisco Conditional Release Program and community outpatient treatment programs, business as usual in the future without some, perhaps a great deal of change like my regaining of my sanity! In 2009, it was not safe for me to "think aloud" with my CONREP therapists. My mulling over, pondering, or ruminating—considering a matter fully, like my full disclosures in this letter—required a lengthy hospitalization.

Because I am arguably legally sane, and whereas my hospital, Napa State Hospital, my San Francisco Golden Gate Conditional Release Program, and my community outpatient treatment program are all too liable, legally obligated, and responsible if I, God help me, reoffend, let's have an indemnifying annual sanity hearing and win my complete release after almost forty years! I am sure I will still have the Veterans Affairs helping me with my continuing mental illness.

Sincerely yours,

Mr. Dorian G. Redus

Cc: Dr. Florida, MD
    Ward T-15
    Psychiatrist

Mr. Dorian G. Redus
Ward T-15
Napa State Hospital
2100 Napa Vallejo Hwy.
Napa, CA 94558
(707)252-9988

                                    Thursday, December 16, 2010

Cheryl H. Arkansas
Attorney-at-Law
214 Duboce Avenue
San Francisco, CA 94103
(415)431-0425

Re: Dorian G. Redus PC 1026.

Please find enclosed: one copy of the Monday, April 12, 2010, letter to the Honorable Judge Wyoming.

Dear Attorney:

    My excellent letters do not even adduce sanity without this photocopy of my photocopy of my former San Francisco Conditional Release Program's April 12, 2010, letter to the Honorable Judge Wyoming. How could you have lost it?
    Did you lose your first photocopy perfidiously in a deliberate breach of our hopes and our faith in my high-profile sanity? If you, my attorney-at-law, are insidiously and deleteriously losing our necessary desideratum documents, and we are not even in a possibly high-profile fight in court fighting for my sanity, then I have reasonable trust issues with you as my attorney-at-law, Cheryl H. Arkansas. Perhaps you are purposely losing our case for my sanity deliberately and on purpose and intentionally. I need the capable, helpful, resourceful, empathetic, and winning attorney-at-law that I know you to be. You can do more for me than read and understand

my important San Francisco Superior Court documents in my PC 187. We can win my sanity hearing in Honorable Judge Wyoming's courtroom. Please identify with me, and please understand my situation, feelings, and above all my altruistic motive: winning a sanity hearing in 2011 without criminal obstacles in programs and hospitals stopping me.

*If you were just careless, if you were just accidental and unusually unprofessional regarding one lost letter one time, then I offer you my sincere, necessary, and most appropriate apology for giving you a piece of my angst and almost disgusted mind—for a picayune peccadillo.*

My sanity is not run-of-the-mill, commonplace, or quotidian to me. I need your commitment to winning my sanity hearing in Honorable Judge Wyoming's San Francisco Superior Court. We just cannot go to court without the April 12, 2010, letter to the Honorable Judge Wyoming. That would be court without the corpus delicti—material evidence—showing my adversarial friends at Anka Behavioral Health Services outpatient treatment program have been criminal to the bone in my affairs.

I photocopied the copy you sent to me, erased my pencil notes, and then I photocopied the enclosed copy for you and posterity. Perhaps you may get us better copies from the San Francisco Conditional Release Program. They usually keep a file copy. You could probably call Dr. May, PhD, who helped write the letter. This whole letter has really been regarding our most important phone talk this year, and the most important thing I said during that talk was "We cannot really do anything until I hear the evaluative response on my pet theories." I very recently requested the responses on November 29, 2010, and on December 3, 2010.

Respectfully submitted,

Mr. Dorian G. Redus

Mr. Dorian G. Redus
Ward T-15
Napa State Hospital
2100 Napa Vallejo Hwy.
Napa, CA 94558
707.252.9988

                                    Thursday, December 16, 2010

Dr. Mexico, PhD
Ward T-15
Psychologist
Napa State Hospital
2100 Napa Vallejo Hwy.
Napa, CA 94558

Dear Doctor:

    I think and I feel to talk about what I did to my infant nephew forty-four years ago, going on more than ten years, is the sexual pervert in my caregiver therapist. Because what I did forty-four years ago was nothing, and it is private. My nephew has changed his name.

Sincerely,

Mr. Dorian G. Redus

Cc: Cory Alabama, CSW, Ward T-15

Mr. Dorian G. Redus
Ward T-15
Napa State Hospital
2100 Napa Vallejo Hwy.
Napa, CA 94558-6234
(707)252-9988

Friday, January 7, 2011

Dr. Mexico, PhD
Ward T-15
Psychologist
Napa State Hospital
2100 Napa-Vallejo Hwy.
Napa, CA 94558-6234

Re: my two sentence Thursday, December 16, 2010, letter to you.

Dear Doctor:

 I am sixty-four years old now. In my late teens in San Francisco, I babysat for the approximately two-and-a-half and three- or four-year-old girls of the Blanch and the attorney Mr. Willie White Jr. family, before and just after the birth of their brother and son Michael White. Everything went very well, and they arranged for me to also babysit at my going rate of fifty cents an hour for the Wesley Rhode Island family. He was a pharmacist. They had two youngsters—young girls. Everything went very well again, and they arranged for me to also babysit at my going rate of fifty cents an hour for the David Washington family. He was an English teacher at City College of San Francisco, and she ran a backyard day care center. They had two youngsters—young boys. Everything went very well again, and they arranged for me to also babysit, at my going rate of fifty cents an hour, for Mr. and Mrs. Mort Iowa and family. They had two youngsters—a young girl and her infant brother. He was an accountant. I had a lot of weekend work. I sat for all four families.

One night at the Iowas, things did not go well. I woke up from a good sleep on their living room couch. I saw a suit or sports coat in the dining room. I put my hand in the coat's pocket, and I stole $200 amount, ten $20 bills. Putting the ten twenties in my pocket, I went to see the Iowas' son, the infant Michael Iowa. He was asleep, so I smelled him. He did not stink, so I let him sleep, and I went up the stars to check on his older sister, who was about eight years old at the time. When I expected her to be asleep, it was the middle of the night, she was okay, but she was awake and naked with no clothing, pajamas, sheets, or blankets on her. And furthermore, she was touching her private part, laying on her back. First, I told her, "Don't do that." And the next second, I pulled her bed clothing (blanket) up comfortably around her and covered her. And then I left to go back down the stairs to the living room. Years later, I returned a like amount of the money, but from the night of the theft, there was no more babysitting for those four families.

Around the time of my trouble at the Iowas' babysitting, perhaps just days or weeks after the incident of the nude eight-year-old, one evening, I went to my older sister's, Mrs. Vivette Whitewell (then Mrs. Vivette Biller's), San Francisco apartment. In a complete surprise, I heard her say her marriage that had produced two beautiful young children was over. Her former husband, Mr. Biller, had been gambling with the rent money.

### *What I did to my nephew in 1967*

Then next, after hearing of the divorce, thinking and feeling the Iowas' eight-year-old daughter might be being molested by one of her parents, I wondered why child molesters molest infants and underage children. I looked for a place in my older sister's apartment to think and to feel just what was occupying my pondering mind that woeful evening in 1967. When I found my infant nephew alone on a bed in a small dimly lit bedroom, I went into the small room with its small closet. I took off my infant nephew's little diaper for no proper reason, and I placed him halfway into the doorway of the small closet. Then I bent over nearer and nearer to his midsection as

I pondered, "Why do child molesters molest infants and underage children?" And I bent more as I pondered, "Is the Iowas' eight-year-old daughter being molested by one or both of her parents?" Then next, as I was bending over my infant nephew, I accidentally touched my upper right lip to his bare belly, almost like I was going to give him "belly pudding" to make him laugh and giggle. After all that, and because of the accidental touch, I felt bad even though I was in no way sexually aroused or prurient. I just felt I had just done something improper that I would never ever forget. Again, in the experience or episode with my nephew, I was never concupiscent. I felt no concupiscence, and I have since had no such problem, before or after, with any other underage person. Back to the night at my sister's in 1967. I put my nephew's diaper back on him after the accidental touch and put him back on the bed, where I had found him just minutes before. Then I felt, as I said, I had done something improper that I would never forget and never tell unless I told him first, and then I left the small room, left the apartment on Haight and Scott Street in San Francisco, and I was on my merry way into the evening.

*Why I bring all this up now*

I told all of this to my team on October 1, 2010. And other times, I told all of this to Dr. Eric Florida, MD, my psychiatrist here since September 15, 2010. The last time I spoke to Dr. Florida on this was Thursday, December 16, 2010. And furthermore, at that time, I shared that there is an unreliable and brief mention of this subject and my family in my December 2, 2010, Ward T-15 WRP, on page 6 of that twenty-seven-page plan. That Wellness Recovery Action Plan also states, "Pedophilia has been ruled out by the treatment team based on no known behavior" on the Ward T-15 WRP's page 5. That Ward T-15 report, along with other unreliable reports, is what Dr. Florida and I spoke of on December 16. Dr. Florida had asked to talk to me, as he does each month or so. I spoke to him of getting an important letter off to my attorney-at-law, Cheryl H. Arkansas, that she said she had lost. It was a photocopy of the Monday, April 12, 2010, letter to the Honorable Judge Wyoming that was used to revoke me off my community outpatient treatment program. It had taken me almost two weeks to find someone willing to photocopy it during this lockdown. I was very grateful and felt very successful. I also brought up that I had recently just solved my current and temporary money problems for January and February 2011 by arraigning for San Francisco State University to send me $1,017.84 they had owed me for some months. However, it seemed to me that Dr. Florida just wanted to talk about what I did to my nephew in 1967. He was not even interested in my two "pet theories" (RCTVU or STS), and he was tacit regarding my having recently sent them off to San Francisco State University for their academic evaluative response on November 29 last year. Therefore, when my Ward T-15 psychiatrist, Dr. Eric Florida, MD, said, "He was going to talk to you," my Ward T-15 psychologist, I thought and I felt this: "I think and I feel to talk about what I did to my infant nephew forty-four years ago, going on more than five years, is the sexual pervert in my caregiver therapist. Because what I did forty-four years ago was nothing, and it is private. Moreover, my nephew has changed his name."

I really do not want to sit through another sex offender group here, exposing and airing all my "dirty laundry" for the next two or three years here, and that would give me an additional indebtedness of over $200,000 to $300,000 due to cost of care billing.

Respectfully submitted,

Mr. Dorian G. Redus

Cc: Cory Alabama, CSW, Ward T-15

Mr. Dorian G. Redus
Ward T-15
Napa State Hospital
2100 Napa Vallejo Hwy.
Napa, CA 94558-6234
(707)252-9988

Tuesday, January 11, 2011

Cheryl H. Arkansas
Attorney-at-Law
214 Duboce Avenue
San Francisco, CA 94103
(415)431-0425

Re: Dorian G. Redus, PC 1026.

Please find enclosed: one poem, one very important letter, and one fifteen-page treatise.

Dear Attorney:

 I am testing the waters, as I am occasionally stressed by the perfidiously perpetrated abject hypocrisy that has inappropriately neglected and ignored my RCTVU and my STS theories. But here you now have both of my two theories in their latest entirety.
 Some time, I hope sooner rather than later, I will send you a more official bound photocopy. However, you may read this copy, and you may photocopy it knowing that I give it my approval and, to me, my deeply appropriate kudos.
 Never forgetting that I am only here for wanting to tell the truth, which only caused the "venial sin" of hearing "vicious voices."
 Thank you for being so helpful and considerate.

Respectfully submitted,

Mr. Dorian Gaylord Redus

Mr. Dorian G. Redus
Ward T-15
Napa State Hospital
2100 Napa Vallejo Hwy.
Napa, CA 94558
(707)252-9988

January 17, 2011

Dr. Wendy May, PhD
Anka Behavioral Health Services
Golden Gate Conditional Release Program
350 Brannan Street, Suite 200
San Francisco, CA 94107
(415) 222-6930

Dear Doctor:

I am aborning! While my two theories (RCTVU and STS) are more and more coming into being and my scientific community is getting underway, please peruse these my following words on my basic mores—the acceptable new custom, new rule, and new guidelines for my forensic (formal debating circle) and family community. My adult self-importance and my adult insecurity together cause my psychology of narcissism. While reading the magazine article, "Narcissism at the Ballot Box," from a longer work by Jennifer Senior in New York, I thought my self-importance came from my work on my two theories (RCTVU and STS). And furthermore, I thought my insecurity came from my always being in the relentless "crosshairs" of my Conditional Release Program. The article I read was on page 52 of the November 19, 2010, issue of the news magazine *The Week*. To rid and free myself of my unwanted narcissistic personality trait, I should first promote myself at San Francisco State University by carefully adducing and elucidating my two heuristic theories (RCTVU and STS) and treat myself like most other people do. And I should, second, rid and free myself of my heavy anchor, Anka Behavioral Health Services Golden

Gate Conditional Release Program, knowing that to continue my narcissism is quite simply "simply unacceptable."

We must water my two heuristic seeds (RCTVU and STS). And moreover, we must put my check and my balancing in my Conditional Release Program in my forensic future. There should be no more "sacred cow" Conditional Release Programs with—carte blanche—complete discretion in my forensic future, advising my courts like they are kangaroo courts. Our CONREP, my Napa State Hospital, and my courts have driven me to apoplexy (my 1990s divorce) one time too many. This should be seen as putative and not jaded! That my CONREPs, my Napa State Hospital stays, and my courts have driven me to apoplexy and all of this should be generally agreeable and acceptable and not callously and cynically considered to be so bizarre as to be unworthy of consideration, treatment, and formal attention. Treating my relationships with all women and all my work on my two aborning theories—RCTVU (relativistic color television universe) theory and STS (space-time sphere) theory—as if they are, all three, a warning sign on a road map to some place someone will be killed must stop—and immediately cease and desist yesterday!

Doctor, I think your apology is appropriate and necessary. I have already apologized to you as it was appropriate and necessary. Whatever ensconced on this ward, Ward T-15, where I first met our director Mr. Christopher Idaho, LCSW, ten years ago, I am a low-key cynosure with a low-key low pay sinecure, looking for my complete sanity and my complete freedom right now. I think and feel that I may be here, in bitter regret, for the rest of my natural life. I also think and feel it is not up to me when I return to San Francisco, no matter what therapists say about it being up to me, when I leave here. I behave like I should never have come to Napa State Hospital, and here I have always behaved like I don't belong here.

In the 1990s, a social worker said I was the "most acutely psychotic" patient she had ever seen admitted to this state hospital. And a psychiatrist, Dr. William Wisconsin, MD, said he medicated me because I was catatonic. Both reports meant something if they were true, and a fortiori, both reports mean something else entirely different because the two reports are both lies that paved the way for me to stay here at Napa State Hospital for seven years, which caused my divorce due to irreconcilable differences in the middle of the seven years. It was the house that Jack built, rather, wrecked. The Napa State Hospital social worker's lie and the Napa State Hospital psychiatrist's lie on a superior court document and in a superior court's sworn testimony respectively are both malpractice. And moreover, then there are you and Mr. Idaho, who both sent me back here and revoked me here as you are both reported on in my (enclosed) Monday, November 15, 2010, letter to my attorney-at-law, Cheryl H. Arkansas.

### To-Do List

1. I have communicated all of this to my attorney-at-law, Cheryl H. Arkansas.
2. I am communicating to you.
3. I hope to communicate with my court.
4. If necessary, I hope to communicate with the press.
5. And if necessary, I hope to communicate with my family, university, and the prosecution.

### One Flew Over

I am a realistic pragmatist. If I do not correctly receive all of what I want, all of what I had will do fine. However, I do have a current pipe dream, and I hope it is not a pie in the sky. I have some needs to buy a nice car in San Francisco, lease an apartment at Parkmerced near San Francisco State University, and eventually take classes for a degree there.

Respectfully submitted,

Mr. Dorian G. Redus

Cc: Cheryl H. Arkansas
    Attorney-at-Law
    14 Duboce Avenue
    San Francisco, CA 94103
    (415) 431-0425

Mr. Dorian G. Redus
Ward T-15
Napa State Hospital
2100 Napa Vallejo Hwy.
Napa, CA 94558
(707)252-9988

Monday, January 17, 2011

Dr. Martin Luther King Jr.
Saint Luke's Catholic Chapel
Napa State Hospital
2100 Napa-Vallejo Hwy.
Napa, CA 94558-6234

Re: my personally felt feeling in the late 1970s, and the mid-1980s, and who should accept full responsibility for it.

Dear Dr. Martin Luther King Jr.:

    I am only testing the waters. However, I impugn all sexual charges. If it is about me, then it is either good or a lie. Full disclosure is causing my feared incorrect conviction for child molestation in the hearts and minds of those who treat and love me. Yes, those who fear, treat, and love me are trying to catch me at a horrendous "child molestation" or at a horrendous philandering homosexual act, and this, their trying to catch me, is making my full disclosures, like this letter, an evil thing.
    In the late 1970s and mid-1980s, I was raped twice. Twice, two of my California state hospital psychiatrists drugged me with dangerous "medications" that gave me only one choice, and that choice was to feel off and queer. Moreover, when a policing public conscious was needed, in tacit secret denial, they profiled me as a homosexual philanderer—*in saecula saeculorum*—for ever and ever, and for all eternity with treacherous travesties of justice, the extent of which I do not know and understand. And the extent of which I may never

know and understand. I do not know who a public conscious will see raping whom, but I am not the evil one. I personally am very persuaded to follow my two theories (RCTVU and STS) rather than following a nonexistent child molestation to further oblivion or finding and following abject psychiatric raping without more compensation. The child molestation and the rapes are a hateful path. Whereas the RCTVU and STS theories are a philosophic path and not a pathological path, and I more love my theories and knowledge than I do hate, to me, those that need to catch or convict me are just plane paranoid. Unfortunately, they are also very, very powerful in my forensic affairs, and they lie (double entendre) above the law.

In the late 1970s and mid-1980s, at Atascadero and at Napa, the abject—objectionable, despicable, miserable, wretched, and contemptible—psychiatrists, Dr. Wiggly at Atascadero State Hospital and Dr. Marshal Arizona at Napa State Hospital, tortured me with their choice of psychotropic drugs. Dr. Wiggly's Prolixin and Dr. Marshal Arizona's Haldol. Once again, in the late 1970s and mid-1980s, two state hospital psychiatrists medicated me until I felt both bad and very awful, and until I was so afflicted with queerness and aborning rage that I performed hundreds of illegal acts of consensual oral copulations on Ward 27 in ASH and on Ward Q 3 and 4 in NSH. The victim, me, and the perpetrators, the psychiatrist's medications, are most ironic and incongruous to normal expectations. However, when I imagine staff at the two hospitals talking about my homosexuality, my sentient blaming imagination imagines a very "teachable moment" to elucidate why it was rape, but powerful and intolerant bigotry seems to preponderate in court where I am only another unheard of quotidian dogface without my doctorate degree.

At Atascadero State Hospital and at Napa State Hospital, if I omit my abject feelings and aborning rage from the drugs, in the late 1970s and mid-1980s, then my attitude and my feelings were genial, polite, and without machination. There, other than the two separate years at both hospitals, I was generally as a well-entertained viewer—all the way not queer or homosexual, just a little too studious from 1974 to 2004.

With more adequate legal services and a more sufficient attorney-at-law defending me and my theories, I justly might not have needed the abject medications or needed so much inpatient treatment.

Finally, I have not been party to a homosexual relationship or even performed a single homosexual act since the mid-1980s.

Thank you for our thoughtful time, any help, and all considerations.

Respectfully submitted,

Dorian Gaylord Redus

Mr. Dorian Gaylord Redus
6801 Mission Street, #202
Daly City, CA 94014
(415)994-4204

C/O Napa State Hospital
Ward Q-9
2100 Napa Vallejo Hwy.
Napa, CA 94558-6293
(707)252-9612 or (707)255-9712

August 5, 1994

Dr. Marshall Arizona
Napa State Hospital, Ward A-2
2100 Napa Vallejo Hwy.
Napa, CA 94558-6293
office (707)253-5701, ward (707)253-5398

Carol Engle, Associate Director Government Relations Joint Commission on Accreditation of Health Care Organizations
One Renaissance Boulevard, Oakbrook Terrace, Illinois 60181
(708)916-5600

Re: Dr. Marshall Arizona's 1983 Q-3 & Q-4 client Mr. Dorian Gaylord Redus.

To Whom It May Concern:

    Generally, an efficacious medication may aid a patient in the management of a disease, may ease or relieve a patient's pain, or cure a patient of a disease. Less generally, if a patient has gonorrhea, penicillin may be indicated for a cure. If a patient has an itch, Benadryl may be indicated to ease or relieve a patient's suffering. And if a patient has schizophrenia or diabetes, medication may be indicated to help manage a diseased patient. However, in some cases, the paradigm for efficacy is more or less experimental.

In six weeks of the summer of 1989, as if reborn, I reentered the City College of San Francisco and passed six units with a 3.5 grade point average. Then that fall of 1989, I passed thirteen units in eighteen weeks with a 3.7 grade point average. The point is that summer and fall of 1989, I was (religiously) ingesting my prescribed 5 mg. of Navane.

To terminate: almost thirty years of student enrollment and (obviously) non-enrollment at City College of San Francisco, the fall of 1991 in eighteen weeks, I passed fourteen units with a 4.0 grade point average. And the spring of 1992 in eighteen weeks, I passed seventeen units with a 3.8 grade point average. Moreover, I graduated. The point is that fall and spring, I was (religiously) ingesting my prescribed 20 mg. of Moban. Although my grade point average was high in the summer of 1989, fall of 1989, fall of 1991, and in the spring of 1992, I made low grades in between my 1989 reentry and my 1992 graduation. In the fall of 1990, 10 mg of Navane so incapacitated me that I could not stay enrolled in my City College classes. I dropped all my City College schoolwork except a one-unit class. Then due to the intoxicating 10 mg. of Navane, I dropped it too. And in the spring of 1991, 7 mg. of Navane was so devastating. In eighteen weeks, I only passed two three-unit classes. And although I had never passed a science class with less than an A, that spring 1991, I made a B in English and a B- in anthropology. The foci are zero units passed incapacitated by 10 mg. Navane and six units passed encumbered by 7 mg. of Navane.

As if the therapist was the rapist, on September 25, 1985, I put $37,464 into my Napa State Hospital trust account. My medication regimen was lowered, and then I was transferred to C Ward where a specious Dr. Oldpoor prescribed 5 mg. of Navane, a drug milder than any of my previous psychotropic medications. The deposit may be verified by phoning the NSH's trust officer Al California (707)253-5637 or reading his April 22, 1994, letter to me. Before I went to C Ward, I was in the milieu of Ward 204, a.k.a. Q-3 and Q-4 on the 4 mg. of Haldol. Even an angry expletive jeer was apt to elicit my (make love, not war) illicit sexual cooperation to sexual concupiscence. Although, others were abused, where rape is an illness, mine was iatrogenic rape. Due to therapeutic(?) medications, I solicited uncontrollably and copulated some hundred other patients, all male. No staff, male or female, just patients, and all the illicit concupiscence was doctor-caused illness. Ten years after the PC 187, murder, 1372, and 1026, not guilty by reason of insanity, hospitalizations began, I had my first and only physical fight. *In general*, I was secluded once in March for fighting with a peer from June 21, 1983, unit 204, Q-4, Napa State Hospital team progress conference note. The peer was someone six foot two. But every morning, I mopped up his urine puddles as my bed was next to his or smelled the stench until noon. Although this peer was sick, he was not the rapist.

The cause of this rape:

Dorian: "Why do I have to take the Haldol?"
Dr. Arizona: "Do you want to get out of the hospital?"
Dorian: "Yes."
Dr. Arizona: "Take your medication."

*The effect:*

Jonathan: "I can't believe it went all the way."
Dorian: "Did you ejaculate?"
Jonathan: "Yes, I came."

Although, the second peer, Jonathan Pinkie, looked like the rapist, Dr. Marshall Arizona is the rapist due to his dereliction of duty and damage to prescribe an efficacious medication. Dr. Marshall Arizona is the rapist.

The law regarding this rape:

"Where the person (Dorian) is prevented from resisting by an intoxicating…substance, or controlled substance." (3) Dorian's sexuality was rape as Haldol, an intoxicating controlled substance, prevented him from resting. Whereas the NSH medical record of June

21, 1983, sent to you March 21 of 1994 affirms I was given 4 mgs. of Haldol on Ward Q-3 and Q-4, where one unit of Haldol equals two and one half units of Navane (4 mg. Haldol=10 mg. Navane), where it is true fall 1990 10 mg. of Navane so incapacitated me that I couldn't even finish one class. A fortiori, as I was physical sodomized in 1983. I was raped! "Where the act is accomplished against the victim's will by threatening to incarcerate…and the victim (Dorian) has a reasonable belief that the perpetrator is a public official," (7) where a psychiatrist on a locked ward of a California mental facility is a public official, a fortiori, as I was physically sodomized in 1983. I was raped!

*West's California Codes 1989 Compact Edition*, St. Paul Minnesota: West Publishing Company (800)328-9352 1989. Sec. 261.

*The evidence:*

Dorian: "Why do I have to take medication?"
Dr. Arizona: "Do you want to get out of the hospital?"

# CORRECTION

This copy of *A Three-Part Discussion* was rewritten on Sunday, July 17, 2011, and it is eighteen pages long, not twenty pages.

A Three-Part Discussion
(1) Recession Velocity, (2) the RCTVU,
and (3) a Space-Time Sphere

Dorian Gaylord Redus

Compiled and Rewritten at 2100 Napa Vallejo Hwy.
Napa, CA 94558-6234
On Sunday, July 17, 2011

# Eureka

# Cosmology

It is from numberless, diverse acts of courage and belief that human history is shaped. Each time a person stands up for an ideal, or acts to improve the lot of others, or strikes out against injustice, he [or she] sends forth a tiny ripple of hope.

—Robert F. Kennedy

# ABSTRACT

The subject in the first part, recession velocity, is the expansion and unfolding of space-time between and separating unmoving galaxies. This document's middle section, the RCTVU (relativistic color television universe), has two focuses or foci: (1) how to conceptualize our relativistic color television or its universe and (2) how to imagine a four-dimensional space-time. The third part of this precis is from a space-time sphere. If a naked singularity is an event horizon entered and exited, then the inside of a naked singularity may loom large, and thus an STS (space-time sphere) may be entered and imagined exited by picturing entering its event horizon and next exiting it through its center singularity womb of space-time, as "cosmic censorship" hides it all.

# RECESSION VELOCITY

Albert Einstein's special relativity equation for adding two velocities states unequivocally neither matter nor information may surpass the velocity of light. For almost one hundred years, observed facts have substantiated astronomer Edwin Hubble's original 1929 observations, proclaiming and proving the universe is a universe of galaxies that recede from one another at a recession velocity intrinsically based on their distance apart. Furthermore, no matter how arcane it may be, Dr. Einstein's work and Dr. Hubble's work may cause the unthinkable non sequitur, that both statements may not be true.

One light-year is the distance light travels in one year—nine trillion, 460 billion kilometers or five trillion, 880 billion miles. One parsec is a unit of astronomical distance equal to 3.26 light-years, and one megaparsec (Mpc) is a unit of astronomical distance equal to one million parsecs. Astronomer Edwin Hubble's law clearly states and elucidates that galaxies 4,001 megaparsecs or more distance apart produce a recession velocity greater than the speed or the velocity of light. Hubble's law is the recession velocity of a galaxy equals Hubble's constant (75 km/s/Mpc) times the distance usually given in Mpc between two galaxies. Or Hubble's law is the recession velocity of a galaxy divided by Hubble's constant (75 km/s/Mpc) equals the separating distance usually given in Mpc. However, we must remember that Albert Einstein's special relativity equation for adding two velocities states unequivocally neither matter nor information may surpass the velocity of light; therefore, special relativity's equation for adding two velocities is a speed limit of recession velocities and a size limit. Thus, it seems we may not link or concatenate megaparsecs of distance infinitely without being at a non sequitur or in error.

However, arcane Einstein's work and Hubble's work may be their work is not specious: seemingly true but false. My work above—the way I put them together—is specious, or seemingly true but false. Nevertheless, I am getting ahead of myself here. So what I am getting at will be gotten to in a moment.

Given the speed or velocity of light is 300,000 km/s, recall Hubble's law is the recession velocity of a galaxy equals Hubble's constant (75 km/s/Mpc) times the distance usually given in Mpc between two galaxies. Or Hubble's law is the recession velocity of a galaxy divided by Hubble's constant (75 km/s/Mpc) equals the separating distance usually given in Mpc.

$$75_{(km/s/Mpc)} \times 4,000_{(Mpc)} = 300,000_{(km/s)}$$

Seventy-five kilometers per second per megaparsec times four thousand megaparsecs equals three hundred thousand kilometers per second or the speed or velocity of light.

Or

$$300,000_{(km/s)} / 75_{(km/s/Mpc)} = 4,000_{(Mpc)}$$

Three hundred thousand kilometers per second divided by seventy-five kilometers per second per megaparsec equals the distance four thousand megaparsecs.

And

$$4,000_{(Mpc)} = 4 \times 10^9_{(parsecs)} = 13.04 \times 10^9_{(Ly)}$$

Four thousand megaparsecs equals four billion parsecs equals thirteen point four billion light-years of age in time and diameter in size.

First, the following *words* of Dr. Mario Livio, an astronomer connected with the Hubble Telescope program, compared to my previous specious words make a paradoxical—*truth* opposed to common sense. When I found Dr. Livio's words on the expansion of the universe, especially between galaxies and clusters of galaxies, I had been looking for the answers they provided me for some thirty years. I believe that my previous words are a question, a test, and the answer follows:

> Second, I would like to clarify that it is only the scale of the universe at large, as expressed by the distances that separate galaxies and clusters of galaxies, that is expanding. The galaxies themselves are not increasing in size, and neither are the solar systems, individual stars, or humans; space is simply unfolding between them. Furthermore, a common misunderstanding is to think of the galaxies as if they are *moving* through some preexisting space. This is not the case. Think of the dots on the surface of the [previously mentioned] balloon. Those dots are not moving at all on the surface (which is the only *space* that exists). Rather, *space itself* (the surface) is stretching, thus increasing the distances between galaxies. Finally on this point, the limit from special relativity that matter cannot move faster than light does not apply to the speeds at which galaxies are separated from each other by the stretching of space. As I explained above, the galaxies are not really moving, and there is no limit on the speed with which space can expand. As an aside, I should note that unlike in the case of moving cars or trains, the doppler effect used to determine the red-shift is also in this case a stretching of the wavelength due to the expansion of space itself. (Mario Livio, *The Accelerating Universe: Infinite Expansion, the Cosmological Constant, and the Beauty of the Cosmos*, 49)

So the actual answer and my new point of view is this: what I thought and felt was just the recession velocity is also the everywhere increasing amount of stretching and unfolding of space that separates unmoving galaxies and unmoving clusters of galaxies in our big bang's Hubble bubble—the part of our universe photographed by the Hubble Telescope.

Let's calmly gawk at and ponder this relativistic speed limiting equation in two forms, prose and formula. When you add Velocity A plus Velocity B divided by one plus left hand parenthesis Velocity A multiplied by Velocity B divided by the speed of light squared right hand parenthesis, the sum of two relativistic velocities will never exceed the velocity of light.

$$Va + Vb = \frac{Va + Vb}{1 + \left(\frac{Va \times Vb}{C^2}\right)}$$

One may think of that special relativity equation for adding two velocities and the following special relativity equation for relative motion and time as two siblings. Albeit there are five of them. Let's calmly gawk at and ponder the following Lorentz transformation—the special relativity equation for describing relative motion and time in two forms, prose and formula. Time slows with velocity; the time in motion is proper, rest, time divided by the radicand, one minus left hand parenthesis the velocity of motion divided by the velocity of light, right hand parenthesis squared.

$$T = \frac{T_0}{\sqrt{1 - \left(\frac{V}{C}\right)^2}}$$

*Example*: Suppose that your friend is moving at 98 percent of the speed of light.

Then,

$$\frac{V}{C} = 0.98$$

So that

$$T = \frac{T_0}{\sqrt{1-(0.98)^2}} = 5T_0$$

Thus, a phenomenon that lasts for one second on a stationary clock is stretched out to 5 seconds on a clock moving at 98% of the speed of light. As measured by your fast-moving friend, a 60-second commercial on your TV will last for five minutes and the minute hand on your clock will take 5 hours to make a complete sweep. (William Kaufmann and Roger Freedman, *Universe*, 5th ed., 592–593).

# THE RCTVU

If you have ever wanted to read Dr. Brian Greene's book *The Elegant Universe* or wanted to read Dr. Stephen W. Hawking's three books—*A Brief History of Time, A Briefer History of Time,* and *The Nature of Space and Time*—then you may like perusing this, my essay, on the fabric of relative motion, mass, energy, length, space-time, and special relativity. In Brian Greene's book, relative motion may be perceived in perceptions of George's and Gracie's cell phone conversation and reception, and at the same time relative motion and "travel time effects" may be perceived in this, my essay, imagined on two color television screens.

Let's see this, first, from George's perspective. Imagine that every hour, on the hour, George recites into his cell phone, "It's twelve o'clock and all is well," "It's one o'clock and all is well," and so forth. Since from his perspective Gracie's clock runs slow, at first blush he thinks that Gracie will receive these messages prior to her clock's reaching the appointed hour. In this way, he concludes, Gracie will have to agree that hers is the slow clock. But then he rethinks it: "Since Gracie is receding from me, the signal I send to her by cell phone must travel even longer distances to reach her. Maybe this additional travel time compensates for the slowness of her clock." George's realization that there are competing effects—the slowness of Gracie's clock vs. the travel time of his signal—inspires him to sit down and quantitatively work out their combined effect. The result he finds is that the travel time effect more than compensates for the slowness of Gracie's clock. He comes to the surprising conclusion that Gracie will receive his signals proclaiming the passing of an hour on his clock after the appointed hour has passed on hers. In fact, since George is aware of Gracie's expertise in physics, he knows that she will take the signal's travel time into account when drawing conclusions about his clock based on his cell phone communications. A little more calculation quantitatively shows that even taking the travel time into account, Gracie's analysis of his signal will lead her to the conclusion that George's clock ticks more slowly than hers.

Exactly the same reasoning applies when we take Gracie's perspective, with her sending out hourly signals to George. At first the slowness of George's clock from her perspective leads her to think that he will receive her hourly messages prior to broadcasting his own. But when she takes into account the ever longer distances her signal must travel to catch George as he recedes into the darkness, she realizes that George will actually receive them after sending out his own. Once again, she realizes that even if George takes the travel time into account, he will conclude from Gracie's cell phone communications that her clock is running slower than his.

So long as neither George nor Gracie accelerates, their perspectives are on precisely equal footing. Even though it seems paradoxical, in this way they both realize that it is perfectly consistent for each to think the others clock is running slow. (Brian Greene, *The Elegant Universe*, 45–46)

One should perceive the slow-motion clock and the "travel time effect's" peculiarity in space-time rates above as in the cell phone reception of George and Gracie in relative motion without acceleration as described and with exacting literalness. One should also perceive or see color television broadcasts, George should send to Gracie, and vice versa, broadcasts Gracie should send to George, in relatively slow motion because of the slow-motion clock and the "travel time effect's" peculiarity in space-time rates. These are both because of a governing speed limit, and that limit is the velocity of light, approximately 186,283 miles per second or about 670 million miles per hour. Therefore (this gets deep), one should also perceive and see in accordance with the physical influence of an equation: Einstein's relativity equation for space-time, as the "travel time effect" is also being considered. However, in spite of the fact that George and Gracie perceive each other's hourly messages—"It is twelve o'clock and all is well," "It is one o'clock and all is well," and so forth, after twelve o'clock and after one o'clock on their own clock—their own perception of their own time is proper and perceived as proper time is perceived here on earth.

Looking and calmly gawking at the following first four of the five beautiful special relativity equations for *m*ass, *e*nergy, *l*ength, and *t*ime, we can (may) spell MELT and find the following: (1) the physics of (see note 2 page 76) *mass's* increase with velocity can be imagined on a color television screen, viewed as a relativistic phenomenon, and (this gets deep) used to see, discover, that it is a manifestation or part of our relativistic color television universe. (2) The physics of *energy's* equivalence with mass can be imagined on a color television screen, viewed as a relativistic phenomenon, and (this gets deep) used to see, discover, that it is a manifestation or part of our relativistic color television universe. (3) The physics of *length's* decrease in the direction of motion with velocity can be imagined on a color television screen, viewed as a relativistic phenomenon, and (this gets deep) used to see, discover, that it is a manifestation or part of our relativistic color television universe. (4) The physics of *time's* slowing with velocity can be imagined on a color television screen, viewed as a relativistic phenomenon, and (this gets deep) used to see, discover, that it is a manifestation or part of our relativistic color television universe. (5) Finally, that the adding of two velocities will never exceed the velocity of light can be imagined on a color television screen, viewed as a relativistic phenomenon, and (this gets deep) used to see, discover, that it is a manifestation or part of our relativistic color television universe. The ubiquitous motif, not ridiculous motif, relativistic color television universe theory and its slow-motion clock and its "travel time effect" may be used to describe all of this until we have the best way to describe our mental energy's relativistic existential universe. Moreover, I know, I think, and I feel these five equations of Albert Einstein are the symbols of chi in the United States of America. And furthermore, the relativistic color television universe is, in essence, the way to (this gets deep) discover American chi aborning in the twenty-first century because, for one thing, television was invented in the United States of America. See also number 10 in the bibliography and note 1 at the end of that section.

Try finding the relativistic color television and the relativistic color television universe in the following two excerpts from page 35 and page 36 respectively of Dr. Alan Guth's book *The Inflationary Universe: The Quest for a New Theory of Cosmic Origins*.

> This basic premise of special relativity, however, seems at first to be nonsensical. Suppose, for example, that I am at rest, and a light beam passes me. The speed would have to be c, the standard speed of light. Suppose, now, that I take off in a spaceship to chase the light beam at 2/3 of the speed of light. Common sense (or Newtonian physics) implies that I would then see the light pulse receding form me at only 1/3 of c. The premise of special relativity, however, holds that I would still measure the recession speed as c. No matter how hard I might try to catch a light beam, I will always see the beam recede at c. (Alan Guth, *The Inflationary Universe: The Quest for a New Theory of Cosmic Origins*, 35)

Cogently, time slows for the fast-moving spaceship chasing the beam of light, so the beam of light always moves, is always seen, at c, the speed of light.

> According to Einstein, the spaceship will appear to us to be shorter than the length that would be measured by observers inside the ship, although the width would be unchanged. The clocks on the ship would appear to us to be running slowly, and the clocks at the back of the ship would look like they were set to a later time than the clocks at the front of the ship, even though they would look synchronized to the crew. (Alan Guth, *The Inflationary Universe: The Quest for a New Theory of Cosmic Origins*, 36)

The phenomena of the preceding quotes above are all due to the relativistic, speed of light, television universe's limitation and constraint on movement, causing the speed of light to be a part cosmic television. See the curious and non-intuitive four dimensions defined in the motif above in the velocity of movement, and you may also imagine four-dimensional space-time.

The (late) Dr. Steven W. Hawking, Lucasian professor of mathematics at the University of Cambridge who, powered by his intelligence, was living on and on despite his terminal medical diagnosis, alleges that: "It is often helpful to think of four coordinates of an event as specifying its position in four-dimensional space-time." Where four-dimensional space-time includes length, width, height, and time inside a fifth dimension, I prefer to avoid Dr. Steven W. Hawking's "It is impossible to imagine a four-dimensional space." The preceding pair of quotes are from page 24 of Dr. Hawking's book *A Brief History of Time*. And in it, Hawking not only states categorically four-dimensional space is unimaginable, but he also states that it is relativistic. This is, of course, where relativistic means velocities great enough to approach the velocity of light. Likewise, one of the quirks of the world's scientific community (in general) is that it also maintains that a four-dimensional space-time continuum is relativistic. Where it is *not* impossible to imagine a four-dimensional space-time, if a space-time continuum is relativistic and four-dimensional, then it is part of the relativistic color television universe. Here, the term *relativistic color television universe* is my own (previously) covert concept. However, if it is developed overtly, it may become a term for the twenty-first century's scientific circles and communities at large. Please use it to gain insight into the esoteric realm of relativistic physics. My hope for this enterprise is that it will make the tool relativistic color television universe a user-friendly and honest intellectual edge for helping the scientific community to easily conceptualize—inter alia, among other things, inter alios, among other people—relativistic space-time.

To begin, visualize a seemingly two-dimensional color television as found in most American homes. Paint a mental picture on the screen, but then notice at the same time the two-dimensional screen is also three-dimensional—one height, two width, and three time (ignoring the dimension of depth, unless it is, in the first place, a 3D TV. One may then accept that electronic color televisions show space-time by imagining a clock or a Timex wristwatch as it accurately keeps space-time in your imagination. Furthermore, one may also accept that electronic color television pictures show space-time and some moving astral bodies and vehicles: planets, stars, galaxies, spaceships, and flying saucers. Then one may envision a flying saucer in a four-dimensional environment. Of course, the saucer will never actually leave the screen. On the other hand, one may imagine it traveling at relativistic speeds—receding from the earth, the solar system, and the galaxy. Finally, if we simply augment our cognizance with a second screen, a crucial element of space-time maybe comprehended. Also, remembering the George and Gracie cell phone excerpt, try this next suggestion sitting in front of your TV or computer tonight. Imagine seeing two color television screen images—one on the earth and the other in an extraterrestrial flying saucer out in space receding from the earth. Of the two imagined television screen images, one or both must have relative nonaccelerating motion to the other one and be approaching c, say 98 percent or 99.9997 percent of the speed of light. Next and equally important, they must both be imagined to be broadcasting an electronic color television signal to the other. George broadcasts cosmic TV to Gracie, and Gracie broadcasts cosmic TV to George.

Since the two broadcasting televisions one on earth and the other in a flying saucer are imagined to be moving relativistically to each other, one should also imagine that their rate of broadcast action on the receiving TV set screens is different, slower, coming from the flying saucer to earth and slower coming from earth to the flying saucer. The idea is to make the perceived paradoxical slow-motion clock's peculiarity in broadcast space-time rate in accord with the physical influence of an equation: Einstein's special relativity equation for space-time when in relative motion as the "travel time effect" is also being considered. Both screen's rate of received broadcast space-time should be imagined slower than a Timex wristwatch accurately keeping un-broadcast proper space-time on earth or in the flying saucer. Imagine seeing the earth's TV broadcasting to the flying saucer in relativistically slower motion, and imagine seeing the saucer's TV broadcasting to the earth in relativistically slower motion.

> These effects are summarized in relativity by saying that when someone watches an object recede away from them, that object will be seen to undergo mass dilation, length changes, and time dilation. (David Bodanis, *E=mc²: A Biography of the World's Most Famous Equation*, 82)

The TV set broadcast viewers on earth will see mass dilation, length changes, and time dilation in the flying saucer's broadcasting to them. The TV set broadcast viewers in the flying saucer will see mass dilation, length changes, and time dilation in the broadcasting to them from earth. This arcane relationship is always there and real. However, it is only obvious at most queer velocities or speeds like 98 percent or 99.9997 percent of the velocity or speed of light! Indeed, most readers will require more than an imagined proof. This will come when mankind acquires a real vehicle or phenomenon that is relativistic (see note 3 in this treatise). Then only a few readers will assume RCTVU: every man's relativity theory is erroneous and utterly ridiculous unrealistic craziness. Even now, some more sapient readers may think and feel this four-dimensional thought experiment is not a specious twenty-first century argument, and they may even understand that the all is not an electronic color television, but the all is a relativistic color television universe. To some with discerning gray matter, the following statement may be an ordinary theoretical fact. The idem here is that the rate of broadcast space-time is different, slower, and it is sometimes so through Einstein's special relativity equation for space-time. Thus, the motif is more powerful as evidence than a mere analogy. It is a mathematical, a relativistic, and a visual reality.

> "The medium is the message."
> —Marshall McLuhan

Scientific experiments have proved that Einstein's special relativity equation for space-time expresses exactly how space-time slows with velocity. Thus, the ubiquitous medium Einstein's equation influences is as (includes) the color TV's electronic message, and the ubiquitous medium Einstein's equation influences is as (includes) the relativistic color television universe. Intrinsically, it is cosmic color television universe with the speed of light limiting and constraining motion.

One, we saw televisions show space-time and "travel time effect." Two, we imagined a space-time difference. And three, I explained this is because relativistic space-time is different and not because of some kind of push button illusion. If one imagines seeing the space-time difference mentally on the two imagined TV screens, then, a fortiori, it is even more likely that one will see this relativistic phenomenon everywhere. Moreover, if we see how to imagine with the following MELT and Velocity A plus Velocity B special relativity equations of Dr. Albert Einstein, then a fortiori, it is even more likely that we will all one day see the RCTVU (relativistic color television universe). The continuum of space-time is where and when it is *not* impossible to imagine or see four-dimensional space-time.

Melting Albert Einstein's enamored equations down to prose and formula, we get:

M—mass increases with velocity. A mass in motion is proper, rest, mass divided by the radicand, one minus left hand parenthesis the velocity of motion divided by the velocity of light right hand parenthesis squared.

$$M = \frac{M_0}{\sqrt{1-\left(\frac{V}{C}\right)^2}}$$

E—energy has an equivalent mass. Energy equals mass times the velocity of light squared.

$$E - MC^2$$

L—length in motion shortens measured in the direction of motion. Length in motion, measured in the direction of motion, is the proper, rest, length multiplied by the radicand, one minus left hand parenthesis the velocity of motion divided by the velocity of light right hand parenthesis squared.

$$L = L_0 \times \sqrt{1-\left(\frac{V}{C}\right)^2}$$

T—time slows with velocity. The time in motion is proper, rest, time divided by the radicand, one minus left hand parenthesis the velocity of motion divided by the velocity of light right hand parenthesis squared.

$$T = \frac{T_0}{\sqrt{1-\left(\frac{V}{C}\right)^2}}$$

Va+Vb

Velocity A plus Velocity B

When you add Velocity A to Velocity B, divided by one plus left hand parenthesis Velocity A multiplied by Velocity B divided by the velocity of light squared right hand parenthesis, the sum of the two velocities will never exceed the velocity of light.

$$Va + Vb = \frac{Va + Vb}{1 + \left(\frac{Va \times Vb}{C^2}\right)}$$

Our relativistic color television universe (RCTVU) must have the smaller scale.

According to string theory, the elementary ingredients of the universe are *not* point particles. Rather, they are tiny, one-dimensional filaments somewhat like infinitely thin rubber bands, vibrating to and fro. But don't let the name fool you: unlike an ordinary piece of string, which is itself composed of molecules and atoms, the strings of string theory are purported to lie deeply within the heart of matter. The theory proposes that *they* are ultra microscopic ingredients making up the particles out of which atoms themselves are made. The strings of string theory are so small—on average they are about as long as the Planck length—that they *appear* point like even when examined with our most powerful equipment 2. (Brian Greene, *The Elegant Universe*, 136)

The smallness of the Planck's constant—which governs the strength of quantum effects—and the intrinsic weakness of the gravitational force team up to yield a result called the *Planck length,* which is small almost beyond imagination: a millionth of a billionth of a billionth of a billionth of a centimeter (ten to the minus thirty third centimeter)… If we were to magnify an atom to the size of the known universe, the Planck length would barely expand to the height of an average tree 2. (Brian Greene, *The Elegant Universe*, 130)

Above the Planck length, our universe may just be a cosmic TV. Therefore, our universe may become a four-dimensional relativistic color television universe above the Planck length. And furthermore, our universe may also become a five-dimensional STS in a black hole's singularity under the Planck length, where it is dense enough. Clearly, we are all inside a Hubble bubble. Clearly we are all part of the RCTVU. And clearly, we all may have come from an STS that is open and expanding but contained by a ubiquitous five-dimensional quantum level. The following is an a priori description of a space-time sphere.

# A SPACE-TIME SPHERE

Where is all of this appropriate? Where there is understanding and tutelage for the proclaiming prose and the mathematics from Dr. Stephen W. Hawking's work *The Nature of Space and Time*.

> One could imagine that after being created, the black holes move far apart into regions without magnetic field. One could then treat each black hole separately as a black hole in asymptotically flat space. (Stephen Hawking, *The Nature of Space and Time*, 56–57)
>
> As in the case of pair creation of black holes, one can describe the spontaneous creation of an exponentially expanding universe. One joins the lower half of the Euclidean four-sphere to the upper half of the Lorentzian hyperboloid (fig. 5.7). Unlike the black hole pair creation, one couldn't say the de Sitter universe was created out of field energy in a preexisting space. Instead, it would quite literally be created out of nothing: not just out of the vacuum, but literally be created out of absolutely nothing at all, because there is nothing outside the universe. (Stephen Hawking, *The Nature of Space and Time*, 85)

Therefore, can one create an inflationary period and the big bang out of nothing? Can one "describe the spontaneous creation of an exponentially expanding" cosmic TV universe out of nothing? And can one create an STS out of absolutely nothing?

## A QUOTIDIAN QUASH: FROM MENTAL HYGIENE TO MENTAL HEALTH

If a naked singularity is an event horizon entered and exited, then the inside of a naked singularity may loom large. The following descriptive excerpts are on a black hole's way to make a big bang or space-time sphere. An STS, like a naked singularity, may be entered and imagined exited picturing, entering, its event horizon and next exiting it through its center singularity.

> In *The Life of the Cosmos*. Lee Smolin… posits that a process of self organization like that of biological evolution shapes the universe, as it develops and eventually reproduces through black holes, each of which may result in a new big bang and a new universe. (Lee Smolin, *The Life of the Cosmos,* back cover)
>
> If time ends, then there is literally nothing more to say. But what if it doesn't? Suppose that the singularity is avoided, and time goes on forever inside of a black hole. What then happens to the star that collapsed to form the black hole? As it is forever beyond the [event] horizon, we can never see what is going on there. But if time does not end, then there is something there, happening. The question is what?
>
> This is very like the question about what happened "before the big bang" in the event that quantum effects allow time to extend indefinitely into the past. There is indeed a very appealing answer to both of these questions, which is that each answers the other. A collapsing star forms a black hole, within which it is compressed to a dense state. The universe began in a similarly very dense state from which it expands. Is it possible that these are one and the same dense state? That is, is it possible that what is beyond the [event] horizon of a black hole is the beginning of another universe? (Lee Smolin, *The Life of the Cosmos*, 87–88)

Ersatz—substitute, artificial, or replacement. Something new is always a phantom, an error, or a queer until it is old hat! Is Lee Smolin's universe in a singularity a specious and ersatz connection? Should I believe it is when I have maintained the same connection? I trust the truth in his theory because I have maintained the same through my own heuristic old hat methods. Neil Alden Armstrong's 1969 moonwalk on the moon discussed in 2012. I came up with Dr. Lee Smolin's idea, and I called it a space-time sphere in 1972. However, Neil Alden Armstrong, the first human to walk on the moon, has probably not asserted his description of that 1969 moonwalk on the moon and felt as refuted and neglected as I felt describing my 1972 space-time sphere before reading Dr. Smolin's work *Life of the Cosmos*.

Intrinsically, my space-time sphere paradigm is our observable universe with its inflation and its big bang confined at a singularity's quantum level. And therefore, our singularity's function is to confine others inside relativistic frames of reference in other inner singularities, that each may hollow out, inflate, and form an independent space-time sphere for its own stellar evolution to connect its own intelligent life to its stars, in addition to confining our outside relativistic frame of reference outside the frames of reference inside other inner singularities. Our universe is, thus, that old hat big bang contained inside our STS. And therefore, the inside of our a priori STS is, by relativistic frames, of reference larger inside than outside.

The laws of physics break down at a singularity, and therefore, I posit, think, and feel the speed of light inside our space-time sphere is lesser than the speed of light outside our space-time sphere. The speeds of light inside and outside of our space-time sphere are not the same ineluctably tied together inside to outside. However, each separate inner and outer, both, speeds of light are the same related (tied) to just its own inside or outside frame of reference. The inside is, therefore, identical to its outside. If the respective velocities of light are related to just their own inner or outer frame of reference, then both of the velocities of light are identical. Again, the two, the inner and the outer speeds of light (are), may be ineluctably the same inside to outside (tied) to their own frames of reference. However, the two, both inner and outer speeds of light (are), may be somehow dynamically differential slower inside, making the inside larger inside than outside, if they are related (considered together) and not related to just their own inner or outer frame of reference.

In this paradigm a priori reality, the velocity of the propagating light inside the space-time sphere actually gives it its internal sizing. As in recession velocity, the first paradoxical part of *A Three-Part Discussion*, the space inside this paradigm a priori space-time sphere unfolds and stretches. Again, the velocity of light actually proportions the vast inner space, and the inside of this STS seemingly expands exponentially and forever, as if it were without an outside to the STS universe. Moreover, the STS womb of space-time—all space-time spheres—are most ineluctably immovable by their mass. However, if you do go over the speed or the velocity of light of or inside an STS and you are also part of an RCTVU (relativistic color television universe), then your movement moves—is a calculated action connected to—the original singularities of the space-time sphere. So it is still ad astra per aspera—to the stars through difficulties—unless we harness or control and direct the force of dark energy.

When I see velocities at and over the velocity of light, c + velocities, affect the stretching and unfolding of our exponentially expanding universe, I like to also realize that according to NASA's Wilkinson Microwave Anisotropic Probe (the WMAP satellite) launched in 2001, the universe is 23.3 percent dark matter, 72.1 percent dark energy, and the part of the universe that I base my work on—ordinary matter—is only 4.6 percent of my universe.

# ADDENDUM

This treatise's last two parts are either a growing cancerous delusion or a discovery of merit. If they are a delusion, then I can attack them with the appropriate erudition, as I have done in my treatise's first of its three parts, "Recession Velocity."

As for me, I am currently encouraged that Dr. Steven W. Hawking's cogent words, "It is impossible to imagine a four-dimensional space," from his 1988 book *A Brief History of Time*, are words on a subject Dr. Hawking "almost" completely omits from his more recent 2005 book, *A Briefer History of Time*. Perhaps the impossible was too difficult for too many of his readers to comprehend, or perhaps he was mistaken.

Moreover, in his 1997 book *The Inflationary Universe: The Quest for a New Theory of Cosmic Origins*, author Dr. Alan H. Guth writes, on page 38, that:

> You must imagine a four-dimensional Euclidean space, and then imagine a sphere in the four-dimensional space. The three-dimensional surface of the sphere is precisely the geometry of Einstein's cosmology. (If you have difficulty visualizing a sphere in four Euclidean dimensions, rest assured that you have a lot of company, including the author)

Adducing further! Eureka! In my second reading of *A Briefer History of Time*, I found the following two sentences on page 141 of its conclusion:

> When we combine quantum mechanics with general relativity, there seems to be a new

possibility that did not arise before: that space and time together might form a finite, *four-dimensional space* (italics added) without singularities or boundaries, like the surface of the earth but with more dimensions. It seems that this idea could explain many of the observed features of the universe, such as its large-scale uniformity and also the smaller-scale departures from homogeneity, including galaxies, stars, and even human beings.

# BIBLIOGRAPHY

Bodanis, David. *E=mc²: A Biography of the World's Most Famous Equation.* New York: Walker, 2000.

Greene, Brian. *The Elegant Universe.* New York: W.W. Norton and Company, 1999.

Guth, Alan. *The Inflationary Universe: The Quest for a New Theory of Cosmic Origins.* Reading, Massachusetts: Perseus Books, 1997.

Hawking, Stephen. *A Brief History of Time.* New York: Bantam Books, 1988.

Hawking, Stephen. *A Briefer History of Time.* New York: Bantam Books, 2005.

Hawking, Stephen. *The Nature of Space and Time.* Princeton, New Jersey: Princeton University Press, 1996.

Kaufmann, William and Roger Freedman. *Universe.* 5$^{th}$ ed. New York: W. H. Freeman and Company, 1999.

Livio, Mario. *The Accelerating Universe: Infinite Expansion, the Cosmological Constant, and the Beauty of the Cosmos.* New York: John Wiley & Sons, Inc., 2000.

Primack, Joel R., and Nancy Ellen Abrams. *The View from the Center of the Universe: Discovering Our Extraordinary Place in the Cosmos.* New York: Riverhead Books, a division of Penguin Group (USA), 2006.

Ratti, Oscar, and Adele Westbrook. *Aikido and the Dynamic Sphere.* Boston, Massachusetts: Tuttle, 2001.

Smolin, Lee. *The Life of the Cosmos.* New York: Oxford University Press, 1997.

## Pamphlet

The special relativity equations are from James E. Bradner and Tamar Y. Susskind's *Theories of Relativity* (1995, West Crescent Avenue, Anaheim, California 92801. Litton Instructional Materials, Inc., a division of Litton Industries).

## Note 1

The first two successful television devices were Russian-born American physicist Vladimir Kosma Zworykin's 1923 iconoscope and American radio engineer Philo Taylor Farnsworth's image dissector tube.

## Note 2

Suppose a super space shuttle is blasting along very close to the speed of light. Under normal circumstances, when that shuttle is going slowly, the fuel energy that's pumped into the engines would just raise its speed. But things are different when the shuttle is right at the very edge of the speed of light. It can't go much faster.

Think of frat boys jammed into a phone booth, their faces squashed hard against the glass walls. Think of a parade balloon, with an air hose pumping into it that can't be turned off. The whole balloon starts swelling, far beyond any size for which it was intended. The same thing would happen to the shuttle. The engines are roaring with energy, but can't raise the shuttle's speed, for nothing goes faster than light. But the energy can't just disappear, either.

As a result, the energy being pumped in gets "squeezed" into becoming mass. Viewed from outside, the solid mass of the shuttle starts to grow. There's only a bit of swelling at first, but as you keep on pouring in energy, the mass will keep on increasing. The shuttle will keep on swelling.

It sounds preposterous, but there's evidence to prove it. If you start to speed up small protons, which have one "unit" of mass when they're standing still, at first they simply move faster and faster, as you'd expect. But then, when they get close to the speed of light, an observer really will see the protons begin to change. It's a regular event at the accelerators outside of Chicago, and at CERN…near Geneva, and everywhere else physicists work. The protons first "swell" to become two units of mass—twice as much as they were at the start—then three units, then on and on, as the power continues to be pumped in. At speeds of 99.9997 percent of "c," the protons end up 430 times bigger than their original size.

What's happening is that energy that's pumped into the protons or in our imagined shuttle has to turn into extra mass. Just as the equation states: that "E" can become "m," and "m" can become "E." (David Bodanis, *E=mc²: A Biography of the World's Most Famous Equation*, 51–52)

## Note 3

People haven't traveled like this yet because our fastest rockets move far more slowly than light, but nature does this sort of experiment with elementary particles all the time. Unstable elementary particles called muons are the main component of the cosmic rays that reach low altitudes where most people live, but they would have decayed high in the atmosphere if their lifetimes were not greatly lengthened by this [time dilation] relativistic effect. These muons, which have a half-life of only 1.52 microseconds ($1.52 \times 10^{-6}$ s), live much longer when they move at nearly the speed of light, exactly as predicted by relativity 9. (William Kaufmann and Roger Freedman, *Universe* 5th ed., 340–341)

# LAST WORD

Finally, the following is not holography to me. The following describes something analogous to a TV screen as is found in most American homes, or the following describes something analogous to a PC screen:

> One idea that has drawn a lot of attention follows from a startling theoretical discovery made by Stephen Hawking at the University of Cambridge, UK, and others in the 1970s: quantum effects in the space around a black hole cause it to emit radiation [Hawking radiation] as if it were hot, even though black holes are supposed to swallow mass and energy, not spit it out.
>
> Furthermore, after decades of analysis and generalization of this argument, many physicists now believe that it applies to any three-dimensional volume, from black holes to empty space: the volume's entire information content can be encoded in its two-dimensional surface. Or to put it another way, the ultimate unified theory of everything should describe our apparently solid three-dimensional world in terms of a lower-dimensional reality. Our Universe would emerge from the theory like a three-dimensional optical image from a two-dimensional hologram. [Here, I feel "hologram" is a misnomer for any relativistic color television universe.] (*Nature*, volume 471, March 17, 2011, page 288 by M. Mitchell Waldrop)

The End

Mr. Dorian G. Redus
2100 Napa Vallejo Hwy.
Napa, CA 94558

Monday, November 29, 2010

Deidre A. Defreese
Senior Student Support Services Specialist
Disability Programs and Resource Center
San Francisco State University
1600 Holloway Avenue
Student Services Bldg. 110
San Francisco, CA 94132
Appt: 415.338.2472
Fax: 415.338.1041

Dear Deidre A. Defreese:

    I was a member of San Francisco State University's Disability Programs and Resource Center where you worked with me in 2009. I had transferred in to SFSU from CCSF as a junior in 2009. When I took two classes (psycholinguistics and statistics), we worked together, for an hour, three or four times. I hope that you remember me.
    I took ill, and I left school. I withdrew from SFSU in late September 2009. I am still in the hospital, and I am writing you now because I hope you can help me, in an important way, with an academic hobby.
    I want you to arrange, consult, tutelage, and evaluate. Please pass this on to a professor who can determine if my pet theory is a valid and useful story. I want something like what I have done in "Recession Velocity," part one of my (enclosed) *A Three-Part Discussion* handbook.
    Perhaps someone in SFSU's physics, astronomy, astrophysics, or cosmology department could lend a much-needed helpful hand.

I can pay them. I want to determine that my work, relativistic color television universe theory, is a valid and useful theory.
    Thank you.

Mr. Dorian G. Redus

Mr. Dorian G. Redus
Ward T-15
Napa State Hospital
2100 Napa Vallejo Hwy.
Napa, CA 94558-6234
(707)252-9988

Monday, January 3, 2011

Deidre A. Defreese
Senior Student Support Services Specialist
Disability Programs and Resource Center
San Francisco State University
1600 Holloway Avenue
Student Services Bldg. 110
San Francisco, CA 94132
Appt: (415)338-2472
Fax: (415)338-1041
(415)338-6356 direct line

Dear Deidre A. Defreese:

 I was a member of San Francisco State University's Disability Programs and Resource Center where you worked with me in 2009. I had transferred in to SFSU from CCSF as a junior in 2009. When I took two classes (psycholinguistics and statistics), we worked together, for an hour, three or four times. I hope that you remember me.
 Please let me know if you received my November 29, 2010, cover letter to you with its enclosed copy of my notes on my pet theory, *A Three-Part Discussion*. It is a cosmological treatise. I have a finished copy that I will send to you as soon as I print it, proofread it, and have it bound to make it look good and scholarly.
 Should you need to, you should reach me by phone between the hours of 6:00 a.m. and 9:00 a.m., 1:30 p.m. and 2:30 p.m., after 4:45 p.m., or by mail at the address above. Thank you.

I also need to find out if I will be hearing from one of San Francisco State University's professors any time soon regarding my treatise, *A Three-Part Discussion*. That is why I am contacting you now.

Again, thank you for your help, time, and your consideration.

Respectfully submitted,

Mr. Dorian Gaylord Redus

Mr. Dorian G. Redus
Ward T-15
Napa State Hospital
2100 Napa Vallejo Hwy.
Napa, CA 94558-6234
(707)252-9988

                              Monday, March 28, 2011

Deidre A. Defreese
Senior Student Support Services Specialist
Disability Programs and Resource Center
San Francisco State University
1600 Holloway Avenue
Student Services Bldg. 110
San Francisco, CA 94132
Appt: (415)338-2472
Fax: (415)338-1041
    (415)338-6356 direct line

Re: some notes that I sent to you late last year, and especially regarding this more scholarly bound rewrite of those notes that I have enclosed.

To Whom It May Concern:

    Although I first wrote to you some months ago, and I have not had any kind of follow-up contact by mail or by phone, I am still keeping my hopes of hearing from you alive. I have no internet service here, and I am very technologically embarrassed in other ways. As a patient here, my only quash fighting option is computer word processing in a program suite new to me, on a 3.5" floppy disk, and printing my work on it in a lab half a mile away in another building. However, this is a path that all staff here may accept, follow, and allow with their full approval and personal approbation. We have their support.

I think, and I feel that sending you my treatise, *A Three-Part Discussion*, in its unfinished notes form on Monday, November 29, 2010, helped me to make this enclosed finished manuscript. Thank you for all your past help, time, and consideration. However, please do bless me further with your insightful institution's evaluative response on this my bound copy of my humble work, *A Three-Part Discussion*.

I obsecrate, beg, that you condescend and bless me with your authoritative evaluative response. When I read over my work, *A Three-Part Discussion*, its RCTVU section seems to have a non sequitur or something queer about how author Dr. Brian Greene's "travel time effect" on pages 6 and 7 is connected to my observations on "space-time slowing" due to time dilation on pages 7 and 9. Please help me by shedding some much-needed light on their connection. Is it that time dilation causes Gracie's clock to run slow to George, on page 6, and time dilation causes George's clock to run slow to Gracie on page 7 in the excerpt from Dr. Brian Greene's book *The Elegant Universe*? Here I am referring to the first, the second, and not the third paragraph of the excerpt.

My own evaluative response regarding this, my new enclosed work, is "under a cloud" of quash, and all I think and feel is that God is within us all, sometimes within one as one in their daily life. And I do not want the devil, even sometimes, in me as me in my daily life, so I obsecrate, beg, for your guidance.

Again, thank you for your help, your time, and your considerations. Have a nice day.

Respectfully submitted,

Mr. Dorian Gaylord Redus

Mr. Dorian G. Redus
Ward T-15
Napa State Hospital
2100 Napa Vallejo Hwy.
Napa, CA 94558
(707)252-9988

                                                  Friday, December 3, 2010

NASA
National Aeronautics and Space Administration
Public Services Office
4800 Oak Grove Drive
Pasadena, CA 91109
(818)354-4321

To Whom It May Concern:

    I have two cosmological theories that I need your evaluative response on. Please read all of what I have sent, and please send me your definitive evaluative response on my ideas. Should you need to, you should reach me by phone between 6:00 a.m. and 9:00 a.m., 1:30 p.m. and 2:30 p.m., after 4:45 p.m., or by mail at the address above.
    Previously, my mental health departments have considered my ideas so bizarre as to be unworthy of consideration, and my ideas have previously been summarily seen as dismissible.
    Again, may I please thank you in advance.

Respectfully submitted,

Mr. Dorian G. Redus

Mr. Dorian Gaylord Redus
Ward T-15
Napa State Hospital
2100 Napa Vallejo Hwy.
Napa, CA 94558-6234
1(707)252-9988

Monday, March 14, 2011

National Aeronautics and Space Administration
Public Services Office
4800 Oak Grove Drive
Pasadena, CA 91109
1(818)354-4321

Re: some notes that I sent to you late last year, and especially regarding this more scholarly bound rewrite of those notes that I have enclosed.

To Whom It May Concern:

    Although I first wrote to you some months ago, and I have not had any kind of follow-up contact by mail or by phone, I am still keeping my hopes of hearing from you alive. I have no internet service here, and I am very technologically embarrassed in other ways. As a patient here, my only quash fighting option is computer word processing in a program suite new to me, on a 3.5" floppy disk, and printing my work on it in a lab half a mile away in another building. However, this is a path that all staff here may accept, follow, and allow with their full approval and personal approbation. We have their support.
    I think and I feel sending you my treatise, *A Three-Part Discussion*, in its unfinished notes form on Friday, December 3, 2010, helped me to make this enclosed finished manuscript. Thank you for your help, time, and consideration. However, please do bless me further with your insightful institution's evaluative response on this my bound copy of my humble work, *A Three-Part Discussion*.

I obsecrate, beg, that you condescend and bless me with your authoritative evaluative response. When I read over my work, *A Three-Part Discussion*, its RCTVU section seems to have a non sequitur or something queer about how author Dr. Brian Greene's "travel time effect" on pages 6 and 7 is connected to my observations on "space-time slowing" due to time dilation on pages 7 and 9. Please help me by shedding some much-needed light on their connection. Is it that time dilation causes Gracie's clock to run slow to George on page 6, and time dilation causes George's clock to run slow to Gracie on page 7 in the excerpt from Dr. Brian Greene's book *The Elegant Universe*? Here I am referring to the first, the second, and not the third paragraph of the excerpt.

My own evaluative response regarding this, my new enclosed work, is "under a cloud" of quash, and all I think and feel is "God is within us all, sometimes within one as one in their daily life, and I do not want the devil, even sometimes, in me as me in my daily life, so I obsecrate, beg, for your guidance."

Again, thank you for your help, your time, and your consideration. Have a nice day.

Respectfully submitted,

Mr. Dorian Gaylord Redus

Mr. Dorian G. Redus
Ward T-15
Napa State Hospital
2100 Napa Vallejo Hwy.
Napa, CA 94558-6234
(707)252-9988

Tuesday, January 11, 2011

Dr. Crusher, PhD Ward
T-14 Psychologist
Napa State Hospital
2100 Napa Vallejo Hwy.
Napa, CA 94558

Dear Doctor:

    This letter is to obsecrate, beg, your help. May I please have a formal letter, from you to me, substantiating your verbal statement that my two theories—RCTVU (relativistic color television universe) and STS (space-time sphere)—are not delusions? As you may recall, on November 5, 2010, visiting me in Ward T-15's courtyard, I recall that you said my work, the *Redus Treatise*, renamed *A Three-Part Discussion*, is not delusional. I just need to document your opinion and then substantiate and prove my work with my newest, more scholarly edition of my treatise, *A Three-Part Discussion*, which I will send to you and my university, San Francisco State University, as soon as I print it, proofread it, and have it bound.
    Thank you for all of your help, time, and considerations.

Respectfully submitted,

Mr. Dorian G. Redus

Mr. Dorian G. Redus
Ward T-15
Napa State Hospital
2100 Napa Vallejo Hwy.
Napa, CA 94558-6234
(707)252-9988

Friday, February 25, 2011

Cheryl H. Arkansas
Attorney-at-Law
214 Duboce Avenue
San Francisco, CA 94103
1(415)431-0425

Dear Attorney:

My recurring dreams start with a happy normal situation or activity, move to a dysfunctional acting out of the dream's drama, and then the original pattern of the dream becomes inane, lacking sense and is fast without mental and sometimes even physical substance. Next, the dream's originally happy normal situation worsens, and the inane dream becomes a nightmare of futility and unfeeling frustration. Finally, in the end, of all the recurring dreams/nightmares, my own empathy with my own nascent curiosity and my utter failure to do all things doable wakes me up, and I meet the dark of the night or the ambient light of my aborning morning curiously.

Therapists camouflage my usually sane, calculated, and careful words and actions in clinics, hospitals, and in programs, behest by my courts, with psychotically queer categories, that I, like an escape artist, square peg in a round cylinder, have to escape. Moreover, my therapists one and all, only following orders, doing this, do lie (double entendre) above the law.

To adduce: "He [Dorian] was re-hospitalized briefly [for nine months] in 1990 for adjustment of his psychotropic medication and stress of his sudden marriage to a woman" (from page 1 of an October

13, 2010, Ward T-14, PC 1026 Court Report by the two Napa State Hospital doctors, Dr. Hameed Nebraska, MD, and Dr. Philip Crusher, PhD, staff psychiatrist and staff psychologist, author respectively). That quote and a raft of other subterfuge and "grim assumptions" of my therapists and my family plagued and condemned my relationship and sudden marriage to my former wife, Mrs. Gillian Redus. How can my court believe my sudden November 19, 1992, marriage, my only marriage, sent me to the hospital "briefly in 1990?" How may I prove that I did not even meet my ex-wife, Mrs. Gillian Redus, until October of 1992? Will my court think these statements are queer or the intransigent status quo?

"During these years, until he committed his crime in 1974, he [Dorian] was under the care of a private psychiatrist [I saw the Department of Veterans Affairs, San Francisco and Oakland, Chief of Mental Hygiene, Dr. Donald Montana, MD, in his VA offices during the years 1970–1974]" (from page 4 of an October 14, 2010, Ward T-14, PC 1026 Court Report by the two Napa State Hospital doctors Dr. Hameed Nebraska, MD, and Dr. Philip Crusher, PhD, staff psychiatrist and staff psychologist, author respectively. The private psychiatrist, Dr. Donald Montana, MD, was always at a Department of Veterans Affairs office in San Francisco or Oakland, as their Chief of Mental Hygiene. However, it seems, even today, he has protected and very "deep pockets."

## Some History of the Real Reason I Was Re-Hospitalized in 1990

In 1990, I was re-hospitalized for writing the United States senator heading the senate committee on the Department of Veterans Affairs. My San Francisco Conditional Release Program made the classic case that I had decompensated, and that won me a nine-month re-hospitalization that ended in an expensive court fight, which I ultimately won. The senator wrote me. He would be glad to help. Moreover, I afforded a winning witness, the late Dr. Chris Hatcher, PhD, a nationally acclaimed expert on dangerousness, who said I was not a danger in San Francisco Superior Court. At that time, I had written to the senator because I needed to take the Department of Veterans Affairs to court for the malpractice of their San Francisco and Oakland Chief of Mental Hygiene, Dr. Donald Montana, MD.

## Some Important History of Why I Was Re-Hospitalized in 1994

There was nothing unreasonable or negative regarding my sanity, and there was "no problem," whatsoever, with my sanity! There was nothing unreasonable or negative regarding my marriage, and there was "no problem," whatsoever, with my marriage! My wife was just lovely, and I was financially embarrassed. We were very much in love with each other. However, I paid my attorney $6,000, and that left me unable to put adequate food on my marital table. At that time in 1994, my San Francisco Conditional Release Program "repressed its disagreements" and difficulties with me, and I "expressed my disagreements" with them. Assertively, all I needed was more money to live on and to pay my mounting marital bills and debts. In 1994, Gillian was the most erudite, but we were both finding our own erudite paths, as I continually failed at finding work.

Then at my conditional release program social worker Harry Georgia's last home visit in June 20, 1994, the very day/night before their heavy-handed shriving sic, it was, at first, all intransigent business as usual and a very pleasant status quo. The apartment was immaculately clean, and everyone was all smiles at everyone else until the next morning, June 21, 1994, when I told the Forensic Health Care Conditional Release Program, "I have not had any of my psychotropic medications for about one whole year." Upon hearing my confessing and penitent words, at absolutistic absolution, my conditional release program suddenly pontificated and sent me back to Napa State Hospital with a court order from a San Francisco Superior Court on June 21, 1994. When I confessed my noncompliance with my prescribed medication, their "repression of their disagreements" and difficulties with me stopped, and my "expressing of my well documented raft of disagreements" with them was henceforth and forevermore in vain. "Expressing my disagreements" with their queering measures only made things more pejorative, and even my loving wife's stalwart efforts to save our relationship were summarily and immediately dismissed without discussion, attention to our food problems, attention to formalities like the fact and the truth, and without any compensating attention to my financial difficulties from my fighting my San Francisco Conditional Release Program and its queer lack of commiseration with my aforementioned pecuniary marital problems.

When my court hearings had decompensated and become very dysfunctional, see appellant attorney Mark Christensen's (enclosed) April 1995 Appellant Court brief. Those treacherous travesties of justice, San Francisco's Superior Court proceedings, decided that my therapists at Forensic Health Care San Francisco Conditional Release Program had won their classic case, that I had decompensated.

At closing this my long series of many letters, a treatise, and a brief, although I have always awakened before going too queer in my nightmares, in two of my latest and current recurring dreams, I was fast losing my innate ability to un-queer myself in my nightmares at sixty-four years of age. To escape Napa State Hospital's latest and current queering milieu, I have on two recent occasions awakened just in the nick of time to stop a nightmare's homosexual act. Those two acts were almost beyond my innate ability to stop them. I am not a homosexual, however, within this milieu, it is according to my dreams, harder and harder for me to, in my heart of hearts, rest assured, that I am heterosexual.

Maybe my court should wake up just in the nick of time and declare me legally sane in a paragon San Francisco Superior Court. Will I see that this year? I think, I feel, and I know that my court officers should act now by assuring me that my legal sanity has been miraculously restored as I request on page 6 of my six-page November 15, 2010, letter to my (public defender) attorney-at-law, Cheryl H. Arkansas.

The aforementioned all clearly explained, in further closing comment, I define "quotidian quash" as: the commonplace squashing, silencing, or annulment of disagreements that are clearly explained and asserted.

Thank you for your time and your commiserations.

Respectfully submitted,

Mr. Dorian Gaylord Redus

## COUNTY of SAN FRANCISCO
SAN FRANCISCO, CALIFORNIA

**LICENSE AND CERTIFICATE OF MARRIAGE** — LOCAL REGISTRATION # 3779

|  |  |
|---|---|
| Name of Groom (First, Middle, Last) | DORIAN GAYLORD REDUS |
| Date of Birth | MAY 19, 1946 |
| Residence | 3823 JUDAH ST, SAN FRANCISCO, 94122, SAN FRANCISCO |
| State of Birth | TEXAS |
| Number of Previous Marriages | 0 |
| Usual Occupation | DISABLED STUDENT |
| Usual Kind of Business | COLLEGE |
| Education — Years Completed | 14 |
| Full Name of Father | CALUB RALEIGH REDUS |
| State of Birth (Father) | TEXAS |
| Full Maiden Name of Mother | VIVIENNE FAY STINNETTE |
| State of Birth (Mother) | TEXAS |
| Name of Bride (First, Middle, Last) | GILLIAN MARGARET BARTHOLOMEW |
| Date of Birth | JUN 9, 1970 |
| Residence | 3823 JUDAH STREET, SAN FRANCISCO, 94122, SAN FRANCISCO |
| State of Birth | TRINIDAD |
| Number of Previous Marriages | 0 |
| Usual Occupation | STUDENT |
| Usual Kind of Business | COLLEGE |
| Education — Years Completed | 12 |
| Full Name of Father | DAMIAN BARTHOLOMEW |
| State of Birth (Father) | TRINIDAD |
| Full Maiden Name of Mother | DESIREE JONES |
| State of Birth (Mother) | TRINIDAD |

Signature of Groom: *Dorian G. Redus*
Signature of Bride: *Gillian Bartholomew*

| | |
|---|---|
| Issue Date | NOV 5 1992 |
| License Expires After | FEB 03 1993 |
| License Number | 008519 |
| Name of County Clerk | Bruce R. Jamison |
| County of Issue | SAN FRANCISCO |
| Deputy | FELICIA RUFINO |

Signature of Witness: *Mister Terry*
Address: 400 VAN NESS AVE, SF, CA 94102

Date of Marriage: NOVEMBER 19, 1992
Signature of Person Solemnizing Marriage: *Martel Carde*
Name of Person Solemnizing Marriage: MARTEL CARDE
Religious Denomination: OFFICE MARRIAGE COMMISSIONER
Mailing Address: 400 VAN NESS, SAN FRANCISCO, CA 94102
Date Accepted: NOV 19 1992

27940

STATE OF CALIFORNIA
COUNTY OF SAN FRANCISCO } ss

This is a true and exact reproduction of the document officially registered and placed on file in the office of the SAN FRANCISCO COUNTY RECORDER.

SAN FRANCISCO COUNTY RECORDER

ATTEST:
DATE ISSUED: NOV 19 1992

# A QUOTIDIAN QUASH: FROM MENTAL HYGIENE TO MENTAL HEALTH

Standard Form 507
(Revised August 1954)
Bureau of the Budget
Circular A-32

(Revised 1/72)

**CLINICAL RECORD**

**DATE:** 8/15/74

Report on MENTAL HYGIENE CLINIC - DOROTHY HUGHES
or
Examination/data Prescribed by M-2, Pt. I, 3.03
(Strike out one line) (Specify type of examination or data)

(Sign and date)

**PROGRESS:** See Periodic Report 6/25/74 and David Allen's letter of 5/3/73.

Veteran has been seen in this San Francisco and Oakland Mental Hygiene Clinics since April 16, 1970 on a weekly basis. He was making some adjustment at City College in San Francisco.

On 8/9/74 he notified Dr. Shaskan that he had killed his girl friend, Mary Robinson that morning; that he had given Dr. Shaskan LSD in candy in the past. He was advised to notify his lawyer. On 8/14/74 the newsmedia mentioned that veteran had turned himself in to the police with the body.

The following were notified of the events by telephone:
    Chief of Outpatient Clinic, Oakland
    (Acting Chief Medical Officer)
    Chief of Psychiatry Service, Martinez
    Chief Attorney, Regional Office, San Francisco
    Acting Chief, Mental Hygiene Clinic, San Francisco

**TOTAL CONTACTS**
**THIS ADMISSION:** 87    INDIVIDUAL    MED.    OJT    GROUP 1

**DIAGNOSIS:** Schizophrenic Reaction, paranoid type

**MODIFICATION OF TREATMENT PLAN:** To be determined.

**CURRENT DRUGS:** None

**Prepared by:**

**Approved:**

*Donald A. Shaskan*

DONALD A. SHASKAN, M.D.
SUPERVISING PSYCHIATRIST

Therapist

(Continue on reverse side)

tts.

REDUS, Dorian G.
(Dorn)
DAS:sw
VA, OAKLAND, CA.

C# 2392-55-78
REPORT ON _____ or CONTINUATION OF _____
Distribution: Original not LT File
cc: LT File

COPY

In the Court of Appeal of the State of California
First Appellate District, Division Four

The People of the
State of California

Plaintiff and Respondent,

vs.

Dorian Redus,

Defendant and Appellant.

1 Crim. A068036

San Francisco County
Superior Court
No. 88778

Appeal from the Judgment of the Superior
Court of the State of California for the
County of San Francisco
========================================

Honorable David Garcia, Judge

Appellant's Opening Brief
=========================

Mark L. Christiansen

State Bar #41291
6963 Douglas Boulevard, Suite 296
Granite Bay, California 95746
Telephone (916) 652-0682

Attorney for Appellant
By appointment of the
Court of Appeal under the
First District Appellate Project
Independent Counsel System

In the Court of Appeal of the State of California
First Appellate District, Division Four

The People of the State
Of California

Plaintiff and Respondent,

vs.

Dorian Redus,

Defendant and Appellant.

1 Crim. A068036

San Francisco County
Superior Court
No. 88778

Appeal from the Judgment of the Superior
Court of the State of California for the
County of San Francisco
======================================

Honorable David Garcia, Judge

Appellant's Opening Brief
========================

Mark L. Christiansen

State Bar #41291
6963 Douglas Boulevard, Suite 296
Granite Bay, California 95746
Telephone (916) 652-0682

Attorney for Appellant
By appointment of the
Court of Appeal under the
First District Appellate Project
Independent Counsel System

# TABLE OF CONTENTS

STATEMENT OF APPEALABLE ORDER ............................................. 104

STATEMENT OF THE CASE .......................................................... 105

STATEMENT OF FACTS ............................................................... 108

ARGUMENT .............................................................................. 112

    I.    THERE WAS INSUFFICIENT EVIDENCE TO SUPPORT THE REVOCATION OF OUTPATIENT STATUS ..... 112

    II.    THE TRIAL COURT ERRED IN FAILING TO INQUIRE INTO THE APPELLANT'S COMPLAINTS ABOUT HIS COUNSEL'S PREPARATION AND PERFORMANCE ............................................................ 115

CONCLUSION ........................................................................... 119

# TABLE OF AUTHORITIES

CASES

| | |
|---|---|
| *People v. Bassett* (1968) 69 Cal.2d 122 | 113 |
| *People v. Bean* (1988) 46 Cal.3d 919 | 113 |
| *People v. Harner* (1989) 213 Cal. App. 3d 1400 | 113 |
| *People v. Marsden* (1970) 2 Cal.3d 118 | 116, 118 |
| *People v. Morris* (1988) 46 Cal.3d 1 | 113, 113 |
| *People v. Redmond* (1969) 71 Cal.2d 745 | 113 |
| *People v. Reyes* (1974) 12 Cal.3d 486 | 113, 113 |
| *People v. Smith* (1993) 6 Cal.4$^{th}$ 684 | 118 |
| *People v. Windham* (1977) 19 Cal.3d 121 | 118 |

CODES

| | |
|---|---|
| Penal Code section 1026.2 | 117 |
| Penal Code section 1237 | 104 |
| Penal Code section 1604 | 113 |
| Penal Code section 1606 | 113, 113, 115, 116 |
| Penal Code section 1608 | 104, 116, 106, 114 |
| Penal Code section 1609 | 114 |
| Penal Code section 1610 | 109, 113, 106 |

MISCELLANEOUS

| | |
|---|---|
| California Rules of Court, rule 13 | 104 |

In the Court of Appeal of the State of California
First Appellate District, Division Four

The People of the State of California

Plaintiff and Respondent,

Vs.

Dorian Redus,

Defendant and Appellant.

1 Crim. A068036

San Francisco Country Superior Court No. 88778

Appeal from the Judgement of the Superior Court of the State of California for the County of San Francisco
==========================================================

Honorable David Garcia, Judge

## STATEMENT OF APPEALABLE ORDER

This is an appeal from an order revoking outpatient status pursuant to Penal Code section 1608 proceedings. The appeal is authorized by Penal Code section 1237[2] as an order made after judgment of commitment for insanity which order removed the defendant from the community and incarcerated him at a state mental hospital, thereby affecting his substantial rights (California Rules of Court, rule 13).

---

[2.] Any reference in this brief to a code section is to the Penal Code and to a rule is to the California Rules of Court, unless otherwise specified or clear from the context.

# STATEMENT OF THE CASE

There is no charging petition. no written order of revocation. and no written disposition on file. (Certificate of Clerk filed April 5, 1995.) Some extrapolation from such record as does exist is necessary to prepare a summary of the case. Reference to the reporter's transcription is also necessary because the clerk's minutes are unreliable due to such matters as showing the defendant present when the content of the proceedings makes it clear he was not. The reporter's transcript itself has its own problems, as for example showing the defendant to be saying something on July 6, 1994, when the court notes the defendant is not present.

On December 2, 1993, outpatient status for Mr. Redus was extended to September 12, 1994. (CT 1.) This status apparently followed a disposition of not guilty by reason of insanity many years ago of a case in which the appellant was charged with a murder which had what was referred to as a "sexual component." (CT 2, RT 7.) Mr. Redus was apparently arrested and returned to the locked environment of Napa State Hospital on June 21, 1994, due to expressions which in the outpatient supervisor's opinion were of violent fantasies involving sexual assaults. (RT 8, 110.) However, the reasons for seeking to terminate outpatient status were alleged violations of the outpatient contract in that Mr. Redus needed to be on medication and had not taken his medication in violation of paragraphs 2 and 9. (RT 15, 16, 108.)

Apparently a report was prepared for the court on July 5, 1994, indicating that violation. (RT 14, 15.) On July 6, 1994, neither the defendant nor the defendant's attorney was present in court, and the matter was continued to July 8, 1994. (RT 7/6/94, CT 2.) On July 8, 1994, Mr. Redus was not present, but his attorney, Mr. Barg was and asked for a hearing on "Wednesday" (the following Wednesday would have been July 13), and the matter was set for July 27, 1994, at 2:00 p.m. (RT 7/8/94, CT 4.)

On July 27, 1994, there was a hearing at which Mr. Redus still was not present. It developed that the district attorney did not have his witness available and asked that the transportation of Mr. Redus be cancelled. which the court had done before this hearing took place. (RT 7/27/94, p. 7.) Mr. Redus apparently would not sign a waiver of appearance but also Mr. Barg indicated he had information from some source not set forth on the record[3] that Mr. Redus did not wish to return. (Id., pp. 7-8.) The deputy district attorney present said the case belonged to Mr. Cling who would be back August 15, 1994. The case was set for August 18, 1994. (Id. p. 8.)

Also dated July 27, 1994, there is a letter from Mr. Redus. a couple of unsigned copies of which appear at CT 5 and CT 6. In this letter he asks to be ordered to court and to appear "in propria persona." (CT 5. 6.)

On August 18, 1994, Deputy District Attorney Cling was still not present, and there is no indication in the reporter's transcript that the defendant was present. Mr. Barg noted the matter was on at the district attorney's request for a continuance. Mr. Barg had prepared an order and it was determined that transportation would take place on August 31, 1994, and a return would take place on September 1, 1994. When asked if Mr. Barg would waive time, Mr. Barg responded that they were beyond the hearing date. (RT 8/18/94, pp. 1-2, CT 7.)

On August 31, 1994, the court noted this was a hearing pursuant to Penal Code section 1606 to revoke outpatient status[4]. (RT 8/31/94, p. 1.) Mr. Jacobs of the San Francisco Conditional Release

---

[3] This does not appear to be from Mr. Redus because Mr. Bang stated in conditional form that "if" that was accurate, he did not know what to do.

[4] During a continued hearing on November 2, 1994, Mr. Redus inquired whether the hearing was pursuant to section 1606 or section 1608, pointing out section 1606 would be an annual review, suggesting the section 1608 proceeding should come first, and nothing that the court reporter's notation—apparently at the August 31, 1994, hearing—that it was pursuant to section 1606 was apparently incorrect (RT 118.) The district attorney replied, eventually, that the proceeding was pursuant to Penal Code sections 1608 and 1610 (RT 11/2/94, pp. 118, 119).

Program testified on direct examination, but the hearing was interrupted because it was time to transport Mr. Redus back to the hospital at Napa. There was a request by the defense that the matter be put over to November 2, 1994, to give enough time to procure a transcript for Mr. Redus to read. The matter was continued to November 2, 1994. (Corrected RT 8/31/94. pp. 17-20.)

On November 2, 1994, examination of Mr. Jacobs concluded and the defendant testified. Following this, the court revoked outpatient status based upon the fact that Mr. Redus had violated the provisions of paragraph 9 that he would comply with the terms of the contract to cooperate with the treatment program by not taking medications (RT 11/2/94. pp. 136-139.) Mr. Redus was remanded to the state hospital. (RT 11/2/94, p. 143.)

# STATEMENT OF FACTS

Larry Jacobs, a licensed family therapist with a master's degree in psychology, was the outpatient supervisor for the appellant for three months preceding June 21, 1994, and had additional acquaintance with the case from discussions at weekly staff meetings. (RT[5] 1-3.) Mr. Redus had generally been very cooperative in regard to his medications. In 1991 Mr. Redus had even voluntarily returned to Napa State Hospital (Napa) for three weeks to have the medications adjusted due to some problems with the side effects. (RT 3.) However, in the ten months before June 21, 1994, unbeknown to the outpatient program, he had discontinued taking the medication. (RT 3-4.)

Mr. Redus had signed an outpatient treatment contract on October 16, 1990. (RT 4.)

On the evening of June 20, 1994, Mr. Jacobs had conducted a home visit at about 8:30 p.m., and at that time had observed no difference in the way in which Mr. Redus presented himself. However, the following morning, about eight-thirty, Mr. Redus had come into the outpatient program office and had spoken with the secretary and another outpatient supervisor and then returned home. (RT 5-6.) Based upon his earlier behavior, Mr. Jacobs called Mr. Redus at his home later that day and asked him to come into the office for further evaluation. (RT 6.) During this telephone conversation, Mr. Jacobs noted that Mr. Redus was highly agitated, angry, vaguely threatening, very difficult to comprehend. Mr. Redus made statements which were loose associations, the theme of which was that he wanted to sue the program and expected a large sum of money upon his arrival. (RT 6.)

Mr. Redus showed up for the appointment on time. In the office interview, Mr. Redus expressed himself in "an outstandingly disen-

---

[5] Hereafter, RT will refer to transcripts of August 31 and November 2, 1994.

combulated phrasing of words but interspersed in this he was saying things like this is dangerous. This is serious. I need to get revoked. I don't need medicines. I am going to prove you guys are wrong. I have not taken my medications for months." (RT 6-7.)

At this point. Mr. Jacobs noted Mr. Redus was "more agitated, angry, blaming, and litigious." Given his history, the program called for an ambulance and the police, and pursuant to Penal Code section 1610 had Mr. Redus transported to Napa State Hospital. (RT 7.) It was the impression of Mr. Jacobs that Mr. Redus was suggesting that he wanted to be hospitalized because his wife would be financially provided for while he was in the hospital and also that Mr. Redus wanted his day in court to prove he did not need to be maintained on psychiatric medications.[6] (RT 7.) Mr. Jacobs also noted that the commitment offense had involved a homicide with a sexual component and Mr. Redus had expressed violent fantasies, raw sexual violence, sexual assaults:[7]

Mr. Redus was in violation of his contract, which Mr. Jacobs felt was still in effect, in that he was not taking his medication as prescribed and was not engaged in effective treatment. (RT 16.) In Mr. Jacobs opinion, Mr. Redus currently required extended inpatient treatment and the program could not safely and effectively treat him in the community. (RT 16-17.) Mr. Redus violated his contract by not taking medication he needed for safe supervision in the community.

---

[6] Sometime after the hospitalization, the staff at Napa received a plastic bag with medications, and about a month before the August 31, 1994, hearing, the program received these, consisting of approximately seven hundred Moltin [sic] ten milligram pills, all marked as prescribed to Mr. Dorian Redus (RT 8–9).

[7] Mr. Jacobs had taken down these remarks, which were made as Mr. Redus was being taken down the stairwell by the police (RT 110). Jacobs had told Redus that Jacobs understood Redus to have said he was not taking his medications, and Redus had replied, "No, I have not taken my meds. I have the bottles in a box, by my PDR [Physician's Desk Reference]. In my library. I don't need meds. I am going to try homosexuality and have it shoved up my ass. I have to get revoked. This is serious. This is dangerous. There is no need for me to have my meds since June. I am going to prove that you guys were wrong about my meds. I was so angry at the hospital, I could bit [sic] the artery out of the doctor. I am not going to talk so angry now. The financial settlement you guys owe me will be years of your guys' toil. They make you a homosexual, and I am in the middle of writing an essay on homosexualizing" (RT 110–111).

Supervision in the community is a matter of degree, and when Mr. Redus became "hyperverbal and more loose and tangential, and when his verbal presentations are peppered with violent fantasies and more intense blame, more litigious ideation, more sexual ideation; when he comes across as vaguely or directly threatening," hospitalization was sought. (RT 108-109.)

Paragraph 9 contained the language that: "I will cooperate with my therapeutic program and understand that unless there is a substantial reason to modify this contract, I will abide by this treatment plan for if [sic] a period of one year from today's date." (RT 106.)

Mr. Jacobs said the intent of that language was an option to renegotiate the contract after one year, not to nullify and void it after one year. (RT 106.)

Mr. Redus interjected remarks from time to time throughout the testimony of Mr. Jacobs[8] and also testified. His position was that at a previous hearing (apparently at a hearing on restoration of sanity), Dr. Korpi had testified he was concerned that the medications would cause brain damage. (RT 112.) Mr. Redus had taken his medications throughout 1993 however much suffering the effects caused in his effort to get out of the program. However, he had to stay in the program because he was on medications and they needed to be sure he continued to take them. As a result, Mr. Redus felt if he stopped and was careful, and with the support of his wife, he would be able to get off the medications and demonstrate he did not need them. (RT 124-125.) He also presented a letter from his wife quoting testimony of Dr. Korpi in July 1990 stating concern the medication was poisoning Mr. Redus and expressing concern Mr. Redus was developing brain cancer. Additionally, the letter noted Mrs. Redus had experienced no problems in living with her husband. (CT 9-10.)

Mr. Redus noted that his quarterly reports showed no decompensation over the period he was not taking medications. (RT 112, 117, 136-137.) The judge agreed that there was no evidence of any difficulty up until June 21, 1994, but the issue was what happened

---

[8] Mr. Redus was not pleased with Mr. Barg as his attorney and had asked that Mr. Barg be relieved.

that day. (RT 137-138.) The judge also pointed out that whether or not Mr. Redus needed his medication was not an issue, the question was whether he was violating the terms of the contract. (RT 139.)

Mr. Redus also explained the events of June 21, 1994. He had gone down to the program office and talked with the receptionist, who apparently told him he could be revoked for a year. He returned an hour later to see Mr. Jacobs or Dr. Korpi, but neither was available. He talked to a Mrs. Avery, and he told her he needed to be taken seriously. He was not talking about anything having to do with sex or violence at the time, rather he said that if he had to arrange to be revoked in order to be taken seriously, to go to court, then he would do so. (RT 115.)

He agreed he was upset that day. He explained he had asked Mr. Jacobs over the telephone if he should drive his car because he did not want to leave it on the street if he was going to be revoked. Mr. Jacobs had said to drive down to the office and it would be okay. Then he found out he was going to be revoked. (RT 111-112.) It was only after he found out he was going to be returned to the hospital where he had been abused in the past and the car was going to be left on the street unprotected that he had violent fantasies. (RT 111-112.) He offered to support this with a tape recording Mr. Jacobs had permitted him to make during the interview which showed no problems until Mr. Jacobs had said Mr. Redus needed to be revoked. (RT 112, 114.) When asked by the judge, Mr. Jacobs denied this. (RT 126.) His words as he was going down the stairs were an effort in which he was trying to be physically cooperative. He felt he had been raped in the hospital, as outlined in a letter he presented and the court accepted. (RT 116.) Mr. Redus noted that in 1985 he had written a commission regarding the sort of effect medications were having on him. (RT 116.) The allegations in that letter were not investigated. (RT 116.) In effect, he felt he was a past and likely future victim of rape. (RT 120-121.)

# ARGUMENT

## I
### There Was Insufficient Evidence to Support The Revocation of Outpatient Status

The evidence showed that Mr. Redus had stopped using his medication without any reported decompensation or other notable effect until June 21, 1994. As a result, the judge did not attempt a determination of whether the medication was needed but rather rested his decision on the fact that stopping the medication was a violation of the outpatient contract. (RT 139.)

That contract was signed in 1990 and contained two clauses which were the subject of the dispute regarding whether it had expired. The contract stated. "Unless there is substantial reason to modify the contract, I will abide by this contract for a period of one year from today's date." (RT 135-136.) It also stated, "This contract may be amended or changed at any time at or by supervisor's discretion." (RT 135.)

The judge explained the contract had a termination date of one year but it was implicit that the contract would be carried over year to year. The judge found that under the terms of the contract, Mr. Redus was continued for outpatient treatment on an annual basis after the contract termination date. The contract required annual pictures to be taken for identification. There was a clause that the terms of the contract could not be changed absent good cause. Mr. Redus was admittedly aware that he was supposed to take his medication. Therefore he was in violation of the terms when he did not do so. (RT 139.)

The general rules of appellate review of claims of insufficient evidence are acknowledged. On review of the judgment, the court must presume in support of the judgment the existence of any facts which the trier of fact might reasonably infer from the evidence. (See, e.g., *People v. Bean* (1988) 46 Cal.3d 919, 934.) Substantial evidence must be of the type which reasonably inspires confidence and is of solid value. (See, e.g., *People v. Reyes* (1974) 12 Cal.3d 486, 497; accord *People v. Morris* (1988) 46 Cal.3d 1, 21; *People v. Bassett* (1968) 69 Cal.2d 122, 139.) A reasonable inference may not be based on suspicion or conjecture. A factual finding must be an inference reasonably drawn from the evidence rather than mere speculation on probabilities unsupported by evidence. (See, e.g., *People v. Morris, supra*, 46 Cal.3d at p. 21; *People v. Reyes, supra*. 12 Cal.3d at p. 500: *People v. Redmond* (1969) 71 Cal.2d 745, 755.)

Upon the release of a person to outpatient status, there must first be a plan submitted setting forth specific terms and conditions to be followed during outpatient status. (Pen. Code § 1604.) "Outpatient status shall be for a period not to exceed one year. At the end of the period of outpatient status approved by the court, the court shall [after notice and hearing]...discharge the person from commitment under appropriate provisions of law, order the person confined to a treatment facility, or renew its approval of outpatient status." (Pen. Code § 1606.)

Thus, the trial judge's assumption that the contract was self-renewing was incorrect. Outpatient status is for a one year period or less. That period expired. The mere fact that Mr. Jacobs or the program was under the misapprehension that, despite the clear language of the contract, it was in some form or fashion self-renewing does not extend the statutory period.

The appellant recognizes that the discharge of the person from commitment is not automatic at the end of the period. (See *People v. Harner* (1989) 213 Cal.App.3d 1400[9].) However, by the same token,

---

[9] The person remains jurisdictionally within the province of the court commitment and is physically to remain on outpatient status unless hospitalized under other provision of law. The hearing is to be held no later than thirty days after the one-year period unless good cause exists (Pen. Code § 1606). Under the provisions of Penal Code section 1610, the person may be confined upon a

confinement at a treatment facility or renewal of outpatient status are similarly not automatic. The trial court here did not find there was a danger but rather that there was a violation of outpatient contract conditions. Those conditions were expressly stated to be for a period of one year. If, as Mr. Jacobs felt, it was understood the conditions could be "renegotiated" during that one year period that was fine; but that would not extend the stated period.

There was no basis for the finding that the contract remained in force because the contract, as well as the one year statutory period, had expired. There was no finding that Mr. Redus was dangerous in the community, and the judge expressly stated that his need for medication was not an issue.

As a result, it is respectfully submitted the basis for revocation was not supported, and the evidence was insufficient or the revocation on this basis was an abuse of discretion.

---

revocation requestion pursuant to Penal Code sections 1608 or 1609 pending the court's decision. However, the director of the program is to supply within one judicial day a declaration specifically stating the behavior or other reason for need and a written request for authorization. There is no such declaration in the record unless it is in the sealed material transmitted to this court. Similarly, there is no record of the written request for revocation, and assuredly, the hearing on that was not held within the fifteen judicial days required by Penal Code section 1608 (Certificate of Clerk). The optimistic view of the majority in *Harner, supra*, that by avoiding a jurisdictional interpretation of the language of section 1606, it would not encourage sloppiness and that the publication of the opinion would be sufficient notice appears to have been overly optimistic.

## II

## THE TRIAL COURT ERRED IN FAILING TO INQUIRE INTO THE APPELLANT'S COMPLAINTS ABOUT HIS COUNSEL'S PREPARATION AND PERFORMANCE

As already noted, Mr. Redus apparently first appeared at the August 31, 1994, hearing. At that hearing, he interjected during the testimony that his attorney had told him the only issue was whether or not he was taking his medication which was not accurate. (RT 9-10.) Mr. Redus was certainly correct if this was indeed a hearing pursuant to Penal Code section 1606, as initially announced. Such a hearing would be to determine whether to terminate the commitment entirely, renew outpatient status, or return Mr. Redus to confinement as a danger. This would import the issue of whether Mr. Redus actually needed to continue on medications, a matter expressly not decided and on which neither party offered evidence other than Mr. Redus testimony and efforts to document his position.

Later in the August 31 hearing, Mr. Redus protested that his attorney would not ask if the contract was an annual contract, and the court asked Mr. Jacobs whether that was so. Mr. Jacobs said that as a matter of recent policy the program had decided it would have the contract be signed or renewed annually, using the future tense, which the court understood to mean that such was not current policy. (RT 15-16.) As discussed in Argument I, the law *requires* an annual review and the program's understanding had been in violation of the law.

At the end of the hearing on August 31, Mr. Redus again protested that his attorney was not doing a good job. The court suggested that the matters of concern to Mr. Redus were not before him, and interrupted Mr. Redus's attempt to clarify his previous statements[10]. The interruption was that the judge agreed he could listen to a defense Mr. Redus did not need to take medications. Mr. Redus response was defense counsel had not consulted with him, and the court said defense counsel had said that he would do so. Mr. Redus said his attorney and he disagreed on using that defense. The matter was continued to November 2, 1994. (RT 18-20.)

Significantly, whatever the merits of his position, Mr. Redus at the August 31st hearing was essentially setting forth a protest which was one of disagreement with his attorney's position. He disagreed regarding the nature of the defense and wanted to raise whether the contract was a one year contract (the defense raised this at the November 2, 1994, hearing) and whether he needed to use medications (which was found not to be a relevant consideration at the November 2, 1994, hearing). Neither of these were meritless on their face.

The November 2, 1994, hearing began with defense counsel stating there was going to be a *Marsden*[11] motion. Defense counsel stated it would be based on facts since the last court appearance. The court refused to hear it as being untimely and because they had a *Marsden* hearing every time the case was called. (RT 103-104.) The appellant interjected he wanted his attorney to establish whether the hearing was a Penal Code section 1606 or Penal Code section 1608 hearing, and the court advised it was brought pursuant to section 1606 but asked the district attorney who said he was not sure, it was just to revoke outpatient status. (RT 104-105.) The hearing continued, and during Mr. Redus's testimony he said it made a big difference since the section 1606 hearing would be an annual review to follow the section 1608 hearing. (RT 118.) Still later, during final discussions, Mr. Redus asked why his attorney had not talked with him about a section 1606 hearing and the right of review for outpatient treatment. The district attorney responded there was a jury trial on restoration in December 1993 which resulted in a finding Mr. Redus had not been restored to sanity and that at that time there had been testimony Mr. Redus was not taking his medications. (RT 141.)

---

[10] These statements were rambling with vague references to earlier proceedings.
[11] *People v. Marsden* (1970) 2 Cal.3d 118.

Mr. Redus inquired about his annual review, and the court said the present hearing was not an annual review hearing. (RT 141-142.) Defense counsel and the court both stated that the hospital would bring an extension petition to extend Mr. Redus commitment, and Mr. Redus said he was due an hearing in September and had wanted to discuss that with his attorney but the judge would not permit him to get another attorney. The court said it had ruled and concluded the proceeding. (RT 142-143.)

From the foregoing, it appears that the judge was well on notice as of the start of the November 2, 1994, hearing that there was a problem with the attorney-client relationship. Although Mr. Redus certainly was no attorney, he had grasped some rather significant distinctions. One of those was that by the time of the November hearing not only the contract had expired but so had the annual review period for outpatient status. While restoration of sanity proceedings would generally preclude a further hearing on that subject until December 1994 (see Pen. Code § 1026.2, subds. (e) and (j)), a hearing on renewal of outpatient status was overdue. Mr. Redus's purpose was to achieve a judicial determination of whether it was necessary for him to continue on what he believed to be dangerous medication. His only refusal to engage in a treatment program was his discontinuance of medication, and in a hearing under section 1606 he would be able to litigate that question rather than simply whether he violated his contract.

Defense counsel, however, had had Mr. Redus examined by a court appointed doctor and declined to present evidence regarding Mr. Redus's need to take medication. (RT 132.) Counsel did argue that the prosecution had the burden of proof and that there was no competent testimony of a need for extended inpatient treatment. (RT 134-135.) However, that is not the same as whether there was a need to take medication.

There are obviously matters which could have been the subject of discussion in terms of the attorney-client breakdown: communications regarding the nature of the hearing and what was or was not relevant, investigation of other means of resolving the perceived problem of what might be presented, the nature of the hearing itself, whether there had been adequate communication regarding the defendant's statements at Mr. Jacob's office, whether there were other items of evidence (such as the letters and tapes the appellant sought to submit), and the like. The appellant was entitled at least to be heard on his reasons for his dissatisfaction with counsel.

The trial judge's dismissal of the *Marsden* concerns on the basis that there had previously been hearings is not supported by the record. There had been protests about things counsel was not doing, but there was never a formal hearing into the nature of the appellant's complaints. The judge was apparently operating on a mistaken impression of the facts. As a result, he abused his discretion.

A "Marsden motion" may be made at any time. *(People v. Smith* (1993) 6 Cal.4$^{th}$ 684.) The motion must be timely. *(People v. Windham* (1977) 19 Cal.3d 121.) However, in this case, the motion was made at the appellant's first opportunity. It was based on matters, whatever they may have been, that had arisen since the first hearing.[12] This was his first opportunity to assert the motion.

It is respectfully submitted the motion should have been heard.

---

[12] One clear example of this was the annual review hearing became due in September.

# CONCLUSION

For the foregoing reasons, it is respectfully requested the judgment (order revoking outpatient status) be reversed.
Dated: April ____, 1995. Respectfully submitted.

        Mark L. Christiansen
        State Bar #41291

        6963 Douglas Boulevard, Suite 296
        Granite Bay, California 95746
        Telephone (916) 652-0682

        Attorney for Appellant
        Dorian Redus

        By appointment of the
        Court of Appeal
        in association with the
        First District Appellate Project

# DECLARATION OF SERVICE

I, the undersigned, declare under penalty of perjury as follows:

I am a citizen of the United States, over the age of 18 years and not a party to the within action; my place of employment and business address is: 6963 Douglas Blvd, Ste. 296, Granite Bay, CA 95746.

On April \_\_\_\_\_, 1995, I served the attached

Appellant's Opening Brief

by placing a true copy thereof in an envelope addressed to the person(s) named below at the address(es) shown, and by sealing and depositing said envelope in the United States Mail at Sacramento County, California, with postage thereon fully prepaid. There is delivery service by United States Mail at each of the places so addressed, for there is regular communication by mail between the place of mailing and each of the place(s) so addressed.

Mr. Dorian Redus
2100 Napa Vallejo Highway
Napa, CA 94558-6293

Attorney General
State of California
50 Fremont Street, #300
San Francisco, CA 94105-2239

First District Appellate Project
730 Harrison Street, Suite 201
San Francisco, CA 94107

                                Clerk of the Court
                        Superior Court, San Francisco
                  Hall of Justice, 850 Bryant Street
                     San Francisco, CA 94103-4667
                          attn: Hon. David A. Garcia

Mr. Ira H. Barg, Dept. Public Defender
1550 Bryant Street
San Francisco, CA 94113

                                  District Attorney
                              San Francisco County
                                 850 Bryant Street
                              San Francisco, CA 94103

I declare under penalty of perjury that the foregoing is true and correct.

Executed on April _____, 1995, at Placer County, California.


_____

Mark Christiansen

The appeal was denied.

Mr. Dorian Gaylord Redus
Napa State Hospital, Ward Q-5 and Ward Q-9
2100 Napa Vallejo Hwy.
Napa, CA 94558-6293
(707)255-9712 and (707)252-9612

To Whom It May Concern:

Mrs. Gillian Bartholomew-Redus visited Mr. Dorian Gaylord Redus during the Monday to Friday Napa State Hospital visiting hours 1:30 pm to 4:30 pm, and holiday, Saturday, Sunday weekend visiting hours 9:00 am to 12:00 pm and 1:30 to 4:30 pm.

1. Monday, June 23, 1994, PM
2. Saturday, June 25, 1994, AM and PM
3. Sunday, June 26, 1994, AM and PM
4. Friday, July 1, 1994, AM and PM
5. Monday, July 4, 1994 AM and PM
6. Saturday, July 9, 1994, AM and PM
7. Sunday, July 10, 1994, AM and PM
8. Saturday, July 16, 1994, AM and PM
9. Sunday, July 17, 1994, PM
10. Saturday, July 23, 1994, PM
11. Sunday, July 24, 1994, PM
12. Tuesday, July 26, 1994, PM
13. Friday, July 29, 1994, PM
14. Tuesday, August 2, 1994, PM
15. Sunday, August 7, 1994, PM
16. Saturday, August 13, 1994, PM
17. Sunday, August 21, 1994, PM
18. Sunday, August 28, 1994, PM
19. Saturday, September 10, 1994, PM
20. Sunday, September 11, 1994, PM
21. Saturday, September 17, 1994, PM
22. Sunday, September 18, 1994, PM
23. Saturday, September 24, 1994, AM and PM

\*\*\*\*\*\*\*\*\*\*\*\*\*\*\*\*\*\*\*\*\*\*\*\*\*\*\*\*\*\*\*\*\*\*\*\*\*\*\*\*\*\*\*\*\*\*\*\*\*\*\*\*\*\*\*\*\*\*\*\*\*\*
24. SATURDAY, OCTOBER 1, 1994, AM, PM, SIX HOURS CONSECUTIVE
\*\*\*\*\*\*\*\*\*\*\*\*\*\*\*\*\*\*\*\*\*\*\*\*\*\*\*\*\*\*\*\*\*\*\*\*\*\*\*\*\*\*\*\*\*\*\*\*\*\*\*\*\*\*\*\*\*\*\*\*\*\*
25. SATURDAY, OCTOBER 8, 1994, PM
26. SATURDAY, OCTOBER 22, 1994, PM
27. FRIDAY, NOVEMBER 11, 1994, PM

\*\*\*\*\*ANNIVERSARY\*\*\*\*\*

28. SATURDAY, NOVEMBER 26, 1994
29. TUESDAY, DECEMBER 6, 1994, PM WITH DR. G. AND DAN G
30. SUNDAY, DECEMBER 18, 1994, PM
31. WEDNESDAY, JANUARY 18, 1995, PM
32. FRIDAY, JANUARY 20, 1995, PM
33. MONDAY, FEBRUARY 13, 1995, PM 4 HRS.
34. SUNDAY, FEBRUARY 19, 1995, PM, 1 HR. 4 MIN. AM AND 2 HR. AND 30 MIN
35. THURSDAY, MARCH 9, 1995, PM—LONG PERFECT VISIT
36. THURSDAY, MARCH 16, 1995, PM, 1 HR. AND 40 MIN.
37. WEDNESDAY, MARCH 29, 1995, PM, 1 HR. 50 MIN.—#1 LITHIUM CARBONATE VISIT
38. FRIDAY, MARCH 31, 1995, PM, 1 HR. 15. MIN.
39. TUESDAY, APRIL 4, 1994, PM
40. TUESDAY, APRIL 25, 1995, PM, 3 HR. 45 MIN.
41. FRIDAY, APRIL 28, 1995, PM
42. MONDAY, MAY 8, 1995
43. FRIDAY, MAY 19, 1995, MY BIRTHDAY, WE WENT OUTSIDE FOR HOURS
44. TUESDAY, MAY 30, 1995, RIBS AND CHIRTTLINS
45. FRIDAY, JUNE 9, 1995, HER BIRTHDAY, BLUE SUNDRESS, 1 HR. 4:30 PM–5:30 PM
46. TUESDAY, JUNE 20, 1995, 3:15 PM, 5:55 PM
47. FRIDAY, JUNE 23, 1995
48. FRIDAY, JUNE 30, 1995

Dorian Gaylord Redus patient

## The Laughing Lady

Through a wormhole time tunnel, one may find her at the
demolished San Francisco's
Play Land at the beach.

She was high above the crowds behind a pane of glass,
laughing incessantly.

She wanted to be heard by all with money for a ticket
to enter her house.

Her parents were craft and childhood memory.

Her patriarchal was wooden slides, a wooden centrifugal disk,
some glass mirrors,
and jets of beguiling air that blew up the skirts of unsuspecting
femme fatales.
Her matriarchal was inside.

Demolished, she is moving through time in the
meditations in her memory.

Furthermore, her temperature is from "chilling out"
to concupiscent.

She wears a friendly sports coat, and her face is painted.

Some of her friends are deceased, and others are alive.

She's a memorable automaton personifying laughter and
how time flies,
tempus fugit, time flying.

If she is the laughing lady at the defunct San Francisco's
Fun House,
then you may find her in a warehouse near the WWII submarine,
the USS *Pampanito*'s
exhibit at pier 35 in Fisherman's Wharf.

By Dorian R., Ward T-15 1-1-11

# CONCLUDING ABSTRACT

On my remand to Atascadero State Hospital under my 1975 PC 1026, I have always believed I should only have done ninety more days or one additional year more at some therapeutic state hospital. However, all my many court officers must first unanimously assume my beliefs. We all know my side: I have said my PC 187, murder, was in self-defense, allowed by my iatrogenic (doctor-caused) delusions and paranoia, and I have said that my two theories (RCTVU and STS) needed to be finished and evaluated before a sane adjudication can even begin. Otherwise, all my forensic medical diagnosing, treatments, and therefore, all my courts are way out of line, suicidal for me, and their value is mostly ironic.

# THE REDUS TREATISE READING AND VOCABULARY TEST

(1)  5 points

According to Dr. Mario Livio, PhD, when describing the expansion of the universe, the galaxies and the clusters of galaxies in the universe are moving or unmoving.
Answer: _____

(2)  5 points

In the *Redus Treatise*, what is the four-letter acronym that may help you remember the five special relativity equations of Albert Einstein?
Answer: _____

(3)  10 points

What are the special relativity equations for time and for adding two velocities?
Answer:

(4)   5 points

Who said, "A collapsing star forms a black hole, within which it is compressed to a dense state. The universe began in a similarly very dense state from which it expands. Is it possible that these are one and the same dense state? That is, is it possible that what is beyond the [event] horizon of the black hole is the beginning of another universe?"

Dr. Lee Smolin, Dr. Brian Greene, Dr. Albert Einstein, or Dr. Edwin Hubble?

Answer: _____

(5)   5 points

Which one of the following four does the *Redus Treatise* assert?

    a.   The all is electronic color television, and the all is the relativistic color television universe.
    b.   The all is not electronic color television, and the all is not relativistic color television universe.
    c.   The all is the electronic color television, but the all is not the relativistic color television universe.
    d.   The all is not just an electronic color television, but the all is the relativistic color television universe.

The answer is _____

(6)   5 points

In the *Redus Treatise*, what two items were analogues to two imagined color television screens?

    a.   Two holograms
    b.   Two cell phones communicating
    c.   Two walkie-talkie radios

Answer: _____

(7) 10 points

Fill in of the missing words.

(1) The physics of _____ increase with velocity can be imagined on a color television screen, viewed as a relativistic phenomenon, and (this gets deep) used to see, discover, that it is a manifestation or part of our relativistic color television universe. (2) The physics of _____ equivalence with mass can be imagined on a color television screen, viewed as a relativistic phenomenon, and (this gets deep) used to see, discover, that it is a manifestation or part of our relativistic color television universe. (3) The physics of _____ decrease in the direction of motion with velocity can be imagined on a color television screen, viewed as a relativistic phenomenon, and (this gets deep) used to see, discover, that it is a manifestation or part of our relativistic color television universe. (4) The physics of _____ slowing with velocity can be imagined on a color television screen, viewed as a relativistic phenomenon, and (this gets deep) used to see, discover, that it is a manifestation or part of our relativistic color television universe. (5) That the physics of the adding of two _____ can never add up to be more than the speed of light can be imagined on a color television screen, viewed as a relativistic phenomenon, and (this gets deep) used to see, discover, that it is a manifestation or part of our relativistic color television universe.

(8) 5 points

The physics of the five Einstein special relativity equations described above can be imagined on a color television screen, viewed as a relativistic phenomenon, and (this gets deep) used to see, discover, that they are a manifestation or part of our relativistic color television universe. But what is the minimum number of imagined color television screens necessary?

    a. One
    b. Two
    c. Three or more
Answer: _____

(9) 5 points

What is the word, from the vocabulary list below, that is a synonym for the word *arcane*?
Answer: _____

(10) 100 points

Vocabulary matching: the numbered words are to the left and the lettered meanings are to the right.

(1) Ineluctable_____      (A) An infinitely small and dense thing
(2) Elucidate_____      (B) Wise and insightful
(3) A fortiori_____      (C) Not to be avoided or changed
(4) Esoteric_____      (D) Seemingly true but in fact false
(5) Sapient_____      (E) The explosive start of the universe

(6) Paradoxical_____    (F) A thing or place with a gravitational field from which nothing, not even light, can escape

(7) Treatise_____    (G) Something that is everywhere at the same time

(8) Galaxy_____    (H) A very large group of stars, on average 100 billion of them

(9) Non sequitur_____    (I) A line that approaches but never meets a curve or is tangent to the curve only at infinity

(10) Specious_____    (J) The distance light travels in one year

(11) Black hole_____    (K) To explain clearly

(12) Asymptote_____    (L) A weird, eccentric, or strange thing

(13) Phenomenon_____    (M) A more definite or higher reasoning or conclusion

(14) Singularity_____    (N) Known to or understood by only few

(15) Big bang_____    (O) An educational method using invention or discovery

(16) Unequivocal_____    (P) Without a doubt clear

(17) Heuristic_____    (Q) A written discussion on a subject

(18) Ubiquitous_____    (R) A statement or something that does not follow logically

(19) Light-year_____    (S) A contradictory truth opposed to common sense

(20) Queer_____    (T) A fact that occurs

(11)   10 points

The *Redus Treatise* includes the following quote from page 130 of Dr. Brian Greene's book, *The Elegant Universe*:

> The Smallness of the Planck's constant—which governs the strength of quantum effects—and the intrinsic weakness of the gravitational force team up to yield a result called the *Planck length*, which is small almost beyond imagination: a millionth of a billionth of a billionth of a billionth of a centimeter (ten to the minus thirty third centimeter)… If we were to magnify an atom to the size of the known universe, the Planck length would barely expand to the height of an average tree.

True or false?
Answer: _____

(12)   10 points

The *Redus Treatise* includes the following quote from page 85 of Dr. Steven W. Hawking's book, *The Nature of Space and Time*:

> Unlike the black hole pair creation, one couldn't say the de Sitter universe was created out of field energy in a preexisting space. Instead, it would quite literally be created out nothing: not just out of the vacuum, but literally be created out of absolutely nothing at all, because there is nothing outside the universe.

True or false?
Answer: _____

## A QUOTIDIAN QUASH: FROM MENTAL HYGIENE TO MENTAL HEALTH

(13)   15 points

The *Redus Treatise* includes a great deal of support for the following quote from page 24 of Dr. Steven W. Hawking's book, *A Brief History of Time*: "It is impossible to imagine a four-dimensional space."
True or false?
Answer: _____

(14)   10 points

The *Redus Treatise* includes the following biblical quote from Luke 6:42:

> Either how canst thou say to thy brother, Brother, let me pull out the mote that is in thine eye, when thou thyself beholdest not the beam that is in thine own eye? Thou hypocrite, cast out first the beam out of thine own eye, and then shalt thou see clearly to pull out the mote that is in thy brother's eye.

True or false?
Answer: _____

Thank you for your time and effort. Good luck!

# THE REDUS TREATISE READING AND VOCABULARY TEST KEY

1. 5 points

    According to Dr. Mario Livio, PhD, when describing the expansion of the universe, the galaxies and the clusters of galaxies in the universe are moving or unmoving.
    Answer: unmoving.

2. 5 points

    In the *Redus Treatise*, what is the four-letter acronym that may help you remember the five special relativity equations of Albert Einstein?
    Answer: MELT

3. 10 points

    What are the special relativity equations for time and for adding two velocities?
    Answer:

$$T = \frac{T_0}{\sqrt{1-\left(\frac{V}{C}\right)^2}}$$

$$Va + Vb = \frac{Va + Vb}{1 + \left(\dfrac{Va \times Vb}{C^2}\right)}$$

4. 5 points

Who said, "A collapsing star forms a black hole, within which it is compressed to a dense state. The universe began in a similarly very dense state from which it expands. Is it possible that these are one and the same dense state? That is, is it possible that what is beyond the [event] horizon of the black hole is the beginning of another universe?"

Dr. Lee Smolin, Dr. Brian Greene, Dr. Albert Einstein, or Dr. Edwin Hubble?

Answer: Dr. Lee Smolin

5. 5 points

Which one of the following four does the *Redus Treatise* assert?

    a. The all is electronic color television, and the all is the relativistic color television universe.
    b. The all is not electronic color television, and the all is not relativistic color television universe.
    c. The all is the electronic color television, but the all is not the relativistic color television universe.
    d. The all is not just an electronic color television, but the all is the relativistic color television universe.

The answer is d: The all is not just an electronic color television, but the all is the relativistic color television universe.

6. 5 points

In the *Redus Treatise*, what two items were analogues to two imagined color television screens?

      a.   Two holograms
      b.   Two cell phones communicating
      c.   Two walkie-talkie radios

Answer: c. Two cell phones communicating.

7. 10 points

Fill in of the missing words.

    The physics of *mass*'s increase with velocity can be imagined on a color television screen, viewed as a relativistic phenomenon, and (this gets deep) used to see, discover, that it is a manifestation or part of our relativistic color television universe. (2) The physics of *energy*'s equivalence with mass can be imagined on a color television screen, viewed as a relativistic phenomenon, and (this gets deep) used to see, discover, that it is a manifestation or part of our relativistic color television universe. (3) The physics of *length*'s decrease in the direction of motion with velocity can be imagined on a color television screen, viewed as a relativistic phenomenon, and (this gets deep) used to see, discover, that it is a manifestation or part of our relativistic color television universe. (4) The physics of *time*'s slowing with velocity can be imagined on a color television screen, viewed as a relativistic phenomenon, and (this gets deep) used to see, discover, that it is a manifestation or part of our relativistic color television universe. (5) That the physics of the adding of two *velocities* can never add up to be more than the speed of light can be imagined on a color television screen, viewed as a relativistic phenomenon, and (this gets deep) used to see, discover, that it is a manifestation or part of our relativistic color television universe.

8. 5 points

The physics of the five Einstein special relativity equations described above can be imagined on a color television screen, viewed as a relativistic phenomenon, and (this gets deep) used to see, discover, that they are a manifestation or part of our relativistic color television universe. But what is the minimum number of imagined color television screens necessary?

      a.  One
      b.  Two
      c.  Three or more
Answer: a. One

9. 5 points

What is the word from the vocabulary list below that is a synonym for the word *arcane*?
Answer: Esoteric

10. 100 points

Vocabulary matching: the numbered words are to the left and the lettered meanings are to the right.

(1) Ineluctable _____(C)    (A) An infinitely small and dense thing
(2) Elucidate _____(K)    (B) Wise and insightful
(3) A Fortiori _____(M)    (C) Not to be avoided or changed

(4) Esoteric _____(N)        (D) Seemingly true but in fact false
(5) Sapient _____(B)         (E) The explosive start of the universe
(6) Paradoxical _____(S)     (F) A thing or place with a gravitational field from which nothing, not even light, can escape
(7) Treatise _____(Q)        (G) Something that is everywhere at the same time
(8) Galaxy _____(H)          (H) A very large group of stars, on average 100 billion of them
(9) Non sequitur _____(R)    (I) A line that approaches but never meets a curve or is tangent to the curve only at infinity
(10) Specious _____(D)       (J) The distance light travels in one year
(11) Black hole _____(F)     (K) To explain clearly
(12) Asymptote _____(I)      (L) A weird, eccentric, or strange thing
(13) Phenomenon _____(T)     (M) A more definite or higher reasoning or conclusion
(14) Singularity _____(A)    (N) Known to or understood by only few
(15) Big bang _____(E)       (O) An educational method using invention or discovery
(16) Unequivocal _____(P)    (P) Without a doubt clear
(17) Heuristic _____(O)      (Q) A written discussion on a subject
(18) Ubiquitous _____(G)     (R) A statement or something that does not follow logically
(19) Light year _____(J)     (S) A contradictory truth opposed to common sense
(20) Queer _____(L)          (T) A fact that occurs

11. 10 points

The *Redus Treatise* includes the following quote from page 130 of Dr. Brian Greene's book, *The Elegant Universe*:

> The Smallness of the Planck's constant—which governs the strength of quantum effects—and the intrinsic weakness of the gravitational force team up to yield a result called the *Planck length*, which is small almost beyond imagination: a millionth of a billionth of a billionth of a billionth of a centimeter (ten to the minus thirty third centimeter)... If we were to magnify an atom to the size of the known universe, the Planck length would barely expand to the height of an average tree.

True or false?
Answer: True

12. 10 points

The *Redus Treatise* includes the following quote from page 85 of Dr. Steven W. Hawking's book, *The Nature of Space and Time*:

> Unlike the black hole pair creation, one couldn't say the de Sitter universe was created out of field energy in a preexisting space. Instead, it would quite literally be created out nothing: not just out of the vacuum, but literally be created out of absolutely nothing at all, because there is nothing outside the universe.

True or false?
Answer: True

13.  15 points

The *Redus Treatise* includes a great deal of support for the following quote from page 24 of Dr. Steven W. Hawking's book, *A Brief History of Time*: "It is impossible to imagine a four-dimensional space."
True or false?
Answer: False

14.  10 points

The *Redus Treatise* includes the following biblical quote from Luke 6:42:

> Either how canst thou say to thy brother, Brother, let me pull out the mote that is in thine eye, when thou thyself beholdest not the beam that is in thine own eye? Thou hypocrite, cast out first the beam out of thine own eye, and then shalt thou see clearly to pull out the mote that is in thy brother's eye.

True or false?
Answer: False

Thank you for your time and effort.

# PART TWO

# PART TWO

Mr. Dorian Gaylord Redus, Ward T-15
Napa State Hospital
2100 Napa-Vallejo Hwy.
Napa, CA 94558-6234
1(707)252-9988

Monday, May 23, 2011

Michael Finney
7 On Your Side
ABC7 Broadcast Center
900 Front Street
San Francisco, CA 94111
1(415)954-7777

Dear Mr. Michael Finney:

 To prevail against intransigent situational injustice, I need your station's "7 On Your Side." Please read all of what I have sent to Michael Finney of "7 On Your Side," and please help me to get my 1988 IQ score from Napa State Hospital's Dr. Kepner, PhD.

Respectfully submitted,

Mr. Dorian Gaylord Redus
Patient

Mr. Dorian Gaylord Redus, Ward T-15
Napa State Hospital
2100 Napa-Vallejo Hwy.
Napa, CA 94558-6234
1(707)252-9988

Monday, May 16, 2011

Christopher A. Idaho, LCSW
Community Program Director
Anka Behavioral Health Services
Golden Gate Conditional Release Program
350 Brannan Street, Suite 200
San Francisco, CA 94107
1(415)222-6930

Dear Christopher A. Idaho:

I think and I feel that you should take me out of CONREP's "crosshairs" and that you should quit being my adversary and trying to convict me. In the 1980s, just before I left Napa under CONREP, for the first time in September 1988, I took an IQ test. As I convalesced then and convalesce now, I would like to have and to read my IQ score and my test's report by Dr. Kepner, PhD, who still works here at Napa State Hospital on Ward T-16 as their psychologist.

Before the test in 1988, Dr. Kepner said I would be given my test results, and he said I would be followed. However, to this day, every time I ask Dr. Kepner for the results and my test score, which I have never received, he makes a balk (like in baseball) and only says, "You are a smart guy," and walks away smilingly. This hospital currently bills me over $150,000.00 a year for cost of care every year I am here, and they feed me here on $3.50 a day.

Please have the doctor (or someone) send me the score and the report. As I recall, the intelligence test I took was a *Wechsler Adult Intelligence Scale*, which may be converted to a conventional intelligence quotient (IQ). If I do not receive the score and the report, then

I will think and I will feel, as I live on, that Napa State Hospital and others have perfidiously reneged again!

I need the insight the promised IQ score will provide me, just like Dr. Kepner did, so I do not draw any prejudiced conclusions.

Respectfully submitted,

Mr. Dorian Gaylord Redus
Patient

Cc: Dr. Kepner, PhD
    Staff Psychologist, Ward T-16

Mr. Dorian Gaylord Redus, Ward T-15
Napa State Hospital
2100 Napa-Vallejo Hwy.
Napa, CA 94558-6234
1(707)252-9988

                                      Monday, May 16, 2011

The Honorable Judge Wyoming
California Superior Court
San Francisco, Dept 15
850 Bryant Street
San Francisco, CA 94103

Re: REDUS, DORIAN                    DOB: 05/19/46
    SC#: *88778*                             *CII: M02858702*
                                               PC 1026

*The Subject:* is necessary communication for the maintenance of my legal sanity.

*Legal Status:* I am a sixty-four-year-old African American male under a PC 187, murder. On August 9, 1974, I stabbed a woman I had known for six years to death. In October 1975, I was found not guilty by reason of insanity and remanded to Atascadero State Hospital. I was transferred to Napa State Hospital in 1982, and I was released on outpatient status in September 1988, under the supervision of CONREP. I believe it was in May of 2010 that you revoked my outpatient status.

Dear Judge Wyoming:

    This (half-inch) document, *Quotidian Quash*, as I call it, is mostly a chronological collection of letters from me to my current hospital therapists and letters to my current public defender attorney-at-law, Cheryl H. Arkansas. However, my November 15, 2010, letter to my current public defender does not treat one important—

post hoc, ergo propter hoc—question. Did my October 2009 nervous breakdown have to do with my ordinary university studies, as CONREP has implied in their April 12, 2010, letter requesting my revocation at the end of the (enclosed) binding? Or was my February to October 2009 nervous breakdown caused by the extraordinary—after this, therefore because of this—"Four issues that have made me angry in the recent past" listed on page 6 of my six-page November 15, 2010, letter to my current public defender, Cheryl H. Arkansas (enclosed) chronologically in the binding?

In other words, should I stay here, ensconced, on Ward T-15, or should I return to my studies in the San Francisco California community? Please read the November letter and let me know.

Respectfully submitted,

Mr. Dorian Gaylord Redus
Patient

Mr. Dorian Gaylord Redus, Ward T-15
Napa State Hospital
2100 Napa-Vallejo Hwy.
Napa, CA 94558-6234
1(707)252-9988

Monday, May 16, 2011

Cheryl H. Arkansas
Attorney-at-Law
214 Duboce Avenue
San Francisco, CA 94103
1(415)431-0425
1(415)255-8631 Fax

Re: REDUS, DORIAN　　　　　　　　　　DOB: 05/19/46
　　SC# 88778　　　　　　　　　　　　　*CII: M02858702*
　　　　　　　　　　　　　　　　　　　　*PC 1026*

*The Subject:* is necessary communication for the maintenance of my legal sanity.

*Legal Status*: I am a sixty-four-year-old African American male under a PC 187, murder. On August 9, 1974, I stabbed a woman I had known for six years to death. In October 1975, I was found not guilty by reason of insanity and remanded to Atascadero State Hospital. I was transferred to Napa State Hospital in 1982, and I was released on outpatient status in September 1988 under the supervision of CONREP. I believe it was on May 5, 2010, that you participated in the formal revocation of my community outpatient treatment status.

Dear Attorney-at-Law:

　　I am still fine, and I am still living on here at this hospital for the criminally and mentally insane. However, I am not criminally or mentally insane, and I'm wondering if you are. Are you just ironic, as in the short concluding abstract near the end of *A Quotidian Quash: From Mental Hygiene to Mental Health 1969–2011?* I sent it to the Honorable Judge Wyoming, California Superior Court, San Francisco, Dept. 15, 850, Bryant Street, San Francisco, CA 94103. I sent it to the San Francisco California Golden Gate Conditional Release Program's Christopher Idaho, LCSW, and I have given it to my Ward T-15 social worker, Cory, Alabama, CSW. Finally, I have telephoned you many times over the last four or five months, but I do not recall us connecting this year since I sent you a copy of *A Three-Part Discussion* on Tuesday, January 11, 2011. Thank you, and have a very nice day.

Respectfully submitted,

Mr. Dorian Gaylord Redus
Patient

Mr. Dorian Gaylord Redus
Napa State Hospital
2100 Napa-Vallejo Hwy.
Napa, CA 94558-6234
1(707)252-9988

Monday, May 30, 2011

The Honorable Judge Wyoming
California Superior Court
San Francisco, Dept 15
850 Bryant Street
San Francisco, CA 94103

Re: REDUS, DORIAN  
    SC#: *88778*

DOB: 05/19/46  
CII: *M02858702*  
PC 1026

*The Subject:* is necessary communication for the maintenance of my legal sanity.

*Legal Status*: I am a sixty-five-year-old African American male under a PC 187, murder. On August 9, 1974, I stabbed a woman I had known for six years to death. In October 1975, I was found not guilty by reason of insanity and remanded to Atascadero State Hospital. Then I was transferred to Napa State Hospital in 1982, and I was released on outpatient status in September 1988, under the auspices of San Francisco's CONREP. I believe it was in May of 2010 that you revoked my outpatient status.

Dear Judge Wyoming:

    I am very sorry. I made a mistake. I recently sent you the wrong unedited copies of three important letters. They had a *secrete* that should have been a *secret*, *through* that should have been a *threw*, an *effected* that should have been a changed etc., and some spacing errors.

Please accept these three edited copies. Thank you.

Respectfully submitted,

Mr. Dorian Gaylord Redus

Mr. Dorian Gaylord Redus, Ward T-15
Napa State Hospital
2100 Napa-Vallejo Hwy.
Napa, CA 94558-6234
1(707)252-9988

Thursday, May 19, 2011

The Honorable Judge Wyoming
California Superior Court
San Francisco, Dept 15
850 Bryant Street
San Francisco, CA 94103

Re: REDUS, DORIAN          DOB: 05/19/46
    SC#: 88778                 CII: M02858702
                                     PC 1026

*The Subject:* is necessary communication for the maintenance of my legal sanity.

*Legal Status*: I am a sixty-five-year-old African American male under a PC 187, murder. On August 9, 1974, I stabbed a woman I had known for six years to death. In October 1975, I was found not guilty by reason of insanity and remanded to Atascadero State Hospital. I was transferred to Napa State Hospital in 1982, and I was released on outpatient status in September 1988, under the supervision of CONREP. I believe it was in May of 2010 that you revoked my outpatient status.

Dear Judge Wyoming:

    I want and need to win my next sanity hearing or trial, and this is my opening statement. The paragraph below is a quote from page 5 of a six-page letter from my San Francisco Golden Gate Conditional Release Program community outpatient treatment program to you, the Honorable Judge Wyoming. The letter is dated April 12, 2010.

## A QUOTIDIAN QUASH: FROM MENTAL HYGIENE TO MENTAL HEALTH

This whole letter below was written in anticipation as a major letter to reply to you if you reply to me after you have perused my bound ninety-eight-page document, *A Quotidian Quash: From Mental Hygiene to Mental Health 1969–2011.*

*3) Anger*

> Mr. Redus is unable to moderate his anger and has exhibited signs of uncontrollable anger and rage during his instant offense. At the time, he believed that his girlfriend/victim was being unfaithful and that she wanted to harm him. In the past, he has gotten into physical altercations with his girlfriend, which has led to him striking her. Mr. Redus was initially in denial about being angry with his daughter for lending his money to someone and not being able to get repaid. He did not admit that he was angry until he was brought before the treatment team (on September 28, 2009) to discuss about his anger which manifested in having intrusive thoughts about hurting others. Although Mr. Redus may recognize some of his warning signs when he gets angry: irritability and agitation, thoughts about violence, his face turns red, he starts becoming delusional, and starts to have homicidal ideation, nevertheless, he was not forthcoming about his intrusive thoughts. He kept his anger to himself and minimized his warning signs for 8 months before finally telling staff. This is an area where he needs to address his issues and focus on being honest with staff. (From page 5 of a six-page letter from my San Francisco Golden Gate Conditional Release Program community outpatient treatment program to you, the Honorable Judge Wyoming, dated Monday, April 12, 2010)

## Anger

CONREP said, "Mr. Redus is unable to moderate his anger and has exhibited signs of uncontrollable anger and rage during his instant offense." That is a ramifying lie in print by liars who lie (double entendre) above the law. Really, I posit, due to my superior range of intelligence, my personal anger management is my forte or strong point.

My CONREP said, "At the time, he believed that his girlfriend/victim was being unfaithful and that she wanted to harm him." After my 1969–1974 "girlfriend/victim," Ms. Edna Ella Robenson, threw me out of our little in-law love cottage across the street from City College of San Francisco, I visited her. And at times, I saw her with another young man. She told me that she wanted to have sex with her visiting young man, student Mr. Lee Lenard. So for me, Edna was freer than she was unfaithful to me. Moreover, within the first few months (of meeting her), Edna became suicidal once and also attacked me with our kitchen knife, which happened many more times during the six years we were in our doctor-advised relationship. Sometimes she attacked me with our kitchen knife, and she also called San Francisco Police to our small apartments and our little backyard in-law love cottage. Furthermore, when I was visiting her just before I took her life on the worst morning of my life, I found her on the telephone, in her sheer off-white bra and panties.

After Edna got off of the telephone, she said, "That was the police. They are getting me a gun." Then Edna said she was going to kill me with the gun—expletive deleted. All that puts a different light on CONREP's "At the time, he believed that his girlfriend/victim was being unfaithful and that she wanted to harm him."

CONREP's casuistry continues with CONREP's "In the past, he has gotten into physical altercations with his girlfriend, which has led to him striking her." Because I thought and felt she was "flirting with death" and needed a fair game, one time, sort of "warning shot across her bow," I struck her one time—once. What does really anger me and get my goat is CONREP's quote:

> Mr. Redus was initially in denial about being angry with his daughter for lending his money to someone and not being able to get repaid. He did not admit that he was angry until he was brought before the treatment team (on September 28, 2009) to discuss about his anger which manifested in having intrusive thoughts about hurting others.

I did not feel anger at my daughter, Ms. Elaine Rose Hawaii, because I more thought and felt I should choose to be protective than I felt I should choose to feel angry with her. Elaine is putting my youngest granddaughter through the University of San Francisco, an auspicious undertaking at an expensive university for a single mother of two. My youngest granddaughter graduates in May 2011, and that is, obviously to me, more important to me than the $24,000 my daughter stole from my personal savings account. Besides all that, to date, my daughter has paid me back, to date, probably $7,500. Besides all of that, I am angry at CONREP's casuistry because their casuistry used my supportive and loving extended family member as a "scapegoat."

> Although Mr. Redus may recognize some of his warning signs when he gets angry: irritability and agitation, thoughts about violence, his face turns red, he starts becoming delusional, and starts to have homicidal ideation, nevertheless, he was not forthcoming about his intrusive thoughts. He kept his anger to himself and minimized his warning signs for 8 months before finally telling staff. This is an area where he needs to address his issues and focus on being honest with staff.

This is subtle, specious, and harmful "scapegoat" reasoning that is, in its own essence, misleading rationalizations. CONREP is unjustly devising their own self-satisfying false reasons for my supposed behavioral anger with my daughter, Ms. Elaine Rose Hawaii. CONREP's "This is an area where he needs to address his issues and focus on being honest with staff" is a statement that makes me recall my *"Four issues that have made me angry in the recent past,"* which CONREP most probably does not even want to talk about. The *"Four issues that have made me angry in the recent past"* are near the end of my November 15, 2010, six-page letter to my current public defender, Cheryl H. Arkansas. It is all in *A Quotidian Quash*. Actually, I am very lucky that Cheryl took a risk, and she sent me the Golden Gate Conditional Release Program community outpatient treatment program letter to you, dated April 12, 2010, as it explained my revocation, and it caused me to write my reply—*A Quotidian Quash: From Mental Hygiene to Mental Health 1969–2011*.

## Lysergic Acid Diethylamide and My Anger

I did not give the 1970s Department of Veterans Affairs Chief of Mental Hygiene LSD until after he had told me to, for some reason unknown to me, "Get a gun," one or two times. Then I did not next surprise my murder victim in my instant offense, Ms. Edna Ella Robenson, with LSD until we ingested the drug, for the first time, six years after I had met her and after about three thousand hours of sex total, from sex three times a day. And that was all after we had both become each other's lover within days of our first meeting. Although, I felt fear from the stealthy way I had tricked first Dr. Donald Montana, MD, and next Ms. Edna Ella Robenson into ingesting the LSD, I only felt anger from it when they were reticent and mute regarding their LSD experiences, as far as I was concerned. The lysergic acid diethylamide experience with Ms. Edna Ella Robenson in 1974, over thirty-five years ago, was, for me, my last known LSD experience.

## Pharmacologic Rape and Institutional Paranoia

Where I was already walking on eggshells because of my ravaging PC 187 murder conviction, Atascadero State Hospital's medical authorities might have better been more careful than to diagnose me, as they were the preponderating authority with the problem of rationalization. The diagnosis was due to, I assume, my private and peacefully pondering on my personal cosmologies (RCTVU and STS). Their mistake was, for me, very depressing and torturous, and one mistake led to another much more serious mistake—pharmacologic rape. See again my January 17, 2011, letter included chronologically in *Quotidian Quash*. The letter is to the late Dr. Martin Luther King Jr., or just recall the two "terribly appropriate" letters on January 17, 2011, in your copy of my document, *Quotidian Quash*. In my mind, the intimate relationship between the two contradistinctions (RCTVU and STS) and the first (in Atascadero State Hospital) and the second (in Napa State Hospital) abject raping caused the politically "strange bedfellows" that have never been adjudicated or even admitted to. Why? When it cannot be, the lacerating and preponderating secret ramifying assumption may unjustly be that LSD caused the rampant homosexuality first at Atascadero and second at Napa. The fact is the pharmacological raping was due to Atascadero State Hospital's Prolixin and Napa State Hospital's Haldol medications. That is the truth and the fact and not more CONREP casuistry.

## Adducing My Favorite Cosmology

Supposedly, former President John Fitzgerald Kennedy said something like "Some people see things that are, and they ask, why? I see things that never were, and I ask why not?" CONREP has written that I am Axis I: 295.70 schizoaffective disorder, bipolar type. My schizoaffective diagnosis may be caused by my faith in my personal cosmologies (RCTVU and STS). My bipolar part of my diagnosis may be caused by my "up-and-down roller coaster" struggles—discovering and proving to myself that my faith in my personal cosmologies (RCTVU and STS) is a good idea, giving me "mood swings."

You passed a judgment affecting my substantial rights intrinsically based on CONREP's casuistry. Please, I obsecrate, beg, that you ask theoretical faculty member Dr. Lee Smolin, PhD, of the Perimeter Institute for Theoretical Physics, Waterloo, Ontario N2L 2Y5, Canada, if my (STS) space-time sphere is a "well-articulated cosmological delusion about a theoretical universe" or a discovery of merit. He is at lsmolin@perimeterinstitute.ca. What do you say about emailing and paying an expert in Canada? Do you think he will answer an email from you? I certainly hope so, as I quoted Dr. Lee Smolin on page 12 of my treatise, *A Three-Part Discussion*. One of his ideas is the same as one of the ideas in my STS theory. It is all in *Quotidian Quash* after the January 17, 2011, letter to the late Dr. Martin Luther King Jr. The STS theory is on pages 12 and 13 of *A Three-Part Discussion*. Regarding my favorite cosmology, I give my previous court officers an improvement needed. I also feel that it is mentionable that whereas dependency is a common denotation of the word *relativity*, the whole tangential truth to the esteemed Dr. Louisiana's statement that I have "unusually strong and developmentally regressed dependency needs" on page 5 of CONREP's April 12, 2010, letter to you requesting that my outpatient treatment be revoked may be that his term "regressed dependency needs" is more a misnomer (tip of the iceberg) that more describes my de facto repressed relativity theory (RCTVU). Whatever the issue before the court is not theoretical, it is that I am sane. Nevertheless, the casuistry regarding the theoretical and the casuistry regarding the raping are both ravaging.

The lysergic acid diethylamide experience with Ms. Edna Ella Robenson in 1974, over thirty-five years ago, was, for me, my last known LSD experience. The two perpetrators of pharmacological homosexual rape were Atascadero's abject psychiatrist Dr. Wiggly, MD, in the early 1980s, and Napa's abject psychiatrist Dr. Marshal Arizona, MD, in the early mid-1980s. And furthermore, California allows both rapists to lie (double entendre) above the law. Really, I posit, due to my superior range of intelligence, my personal anger management is my forte or my strong point. Finally, I want a quality of mind that enables me to face all my dangers, in full command of my faculties, my feelings, and my thoughts. I want my own courage so I may say that I need my court to accept my excellent anger management. I need my court to trust me. And furthermore, I also need my court to decide that my sanity has been restored at a PC 1606, restoration of sanity hearing.

Respectfully submitted,

Mr. Dorian Gaylord Redus
Patient

Mr. Dorian Gaylord Redus, Ward T-15
Napa State Hospital
2100 Napa-Vallejo Hwy.
Napa, CA 94558-6234
1(707)252-9988

Monday, May 23, 2011

The Honorable Judge Wyoming
California Superior Court
San Francisco, Dept 15
850 Bryant Street
San Francisco, CA 94103

Re: REDUS, DORIAN                    DOB: 05/19/46
    SC#: 88778                       CII: M02858702
                                     PC 1026

*The Subject:* is necessary communication for the maintenance of my legal sanity.

*Legal Status:* I am a sixty-five-year-old African American male under a PC 187, murder. On August 9, 1974, I stabbed a woman I had known for six years to death. In October 1975, I was found not guilty by reason of insanity and remanded to Atascadero State Hospital. I was transferred to Napa State Hospital in 1982, and I was released on outpatient status in September 1988, under the supervision of CONREP. I believe it was in May of 2010 that you revoked my outpatient status.

Dear Judge Wyoming:

In spite of the constitutional guarantees of the United State Constitution of America and my own sanity, injustice in my California mental health system has substantially changed my personal life, liberty, and my normal pursuit of happiness. Why has there been de facto mental corruption to disease in my legal affairs from 1969 to 2011? I know it was thirty years ago. Nevertheless, when I am heterosexual, two of my California mental health systems' hospitals raped me with their abject pharmacological ways until I was homosexually raped for a whole year, twice, changing my life, my liberty, and my pursuits of happiness in and out of the hospitals for going on thirty years. My point is that my "court officers," including my hospitals, have been queering me.

When I was actively heterosexual, and shortly before I killed/murdered one Edna Ella Robenson, my 1974 instant offense victim, I hit her to warn her not to take my kindness for weakness, and I also hit her because she was moving in on me for the de facto—kill. I did not hit and kill her because of my rage, my jealousy, or my envy of her. With reasonable envy of visiting City College of San Francisco lecturers in 1972 to 1974, I wanted to tell our whole world that their summer of 1972, KQED color television TV show on San Francisco's Channel 9 should have been entitled *Stellar Evolution: Man's Ascent from the Stars*, not *Stellar Evolution: Man's Descent from the Stars*, as it was entitled. Back then, I was jealously guarding my life from bitter and dangerous people, and one of the people I was jealously guarding my life from and protecting my life from was one Ms. Edna Ella Robenson. At the time of her death, at my hand, on August 9, 1974, Edna was not a chaste woman with an immaculate heart, according to the religious myth. Also, the doctor in attendance and advising the relationship for six years before and at the time of the killing, one Dr. Donald Montana, MD, never publicly attested to anything in open court and before me and God, stopping the gross injustice that has ensued now for forty years. He died disrespecting me with his silence. All of this has been a quotidian quash. When I trusted him, instead of being trustworthy, he feigned respectability and was publicly unjust to me. The Department of Veterans Affairs medical doctor Dr. Donald Montana, MD, was an abject psychiatrist who I found to be grossly inadequate, unable to help me, and a very dangerous man. Both Edna and Donald had an immaculate way of gawking at me and not saying what was really on their minds. The personally envied visiting City College of San Francisco lecturers involved in CCSF's televised summer of 1972, *Stellar Evolution: Man's Descent from the Stars*, were Professor Duckworth's avant-garde astronomical gauntlet, disclaiming my spelling, Mr. Ray Bradbury, Dr. Geoffrey Burbidge, Dr. Edwin E. Salpeter, Dr. P. J. E. Pebbles, Professor J. W. Schof, Dr. Sherwood Washburn, Dr. Melvin Calvin, Dr. Philip Morrison, Dr. Freeman Dyson, and Dr. Bernard M. Oliver of HP Co. And they should have all known it was my *ascent*, not their *descent*, that connects us all in our human evolution, to our cosmology.

## My Temporary Insanity

As Dr. Donald and I ignored her violence, Edna's violence created in me two jeopardizing psychoses. When Dr. Donald ignored my many disadvantageous disagreements with him and Edna, the disagreements resurfaced in a general psychotic paranoia and a specific psychotic delusion. My suspicions and my disagreements caused paranoid conjecture, and my paranoid conjecture (due to my real jeopardy) did produce a delusion that people do not die. Those psychoses became the conjectures: I may only get the much-needed help I need if I take Edna's life, and that if I don't take her life soon, she will surely kill me first. Therefore, in clinical desperation, as a last resort, I temporarily chose the former—to take Edna's life, like I had learned in the United States of America's US Army. In conclusion, all of this information has been on the tip of my tongue and occasionally coming out of my mouth in words for going on forty years. However, all of this information, like the Department of Veterans Affairs late Dr. Donald Montana, has never come to court and been believed!

I said all of that to say this: I am finding myself sane, and I am finding if it is about me, then it is either good or unjust. I commend to you, Your Honor. Thus, do you find with the long dead Edna Ella Robenson that I am not reasonably envious of the gauntlet? And therefore, do you find it necessary to continue my (*un*bloody) crucifixion, as described in my document, *A Quotidian Quash: From*

*Mental Hygiene to Mental Health 1969–2011*? Or do you find me, sans suspiciousness or conjecture, SANITY RESTORED?

Respectfully submitted,

Mr. Dorian Gaylord Redus
Patient

Mr. Dorian Gaylord Redus, Ward T-15
Napa State Hospital
2100 Napa-Vallejo Hwy.
Napa, CA 94558-6234
1(707)252-9988

Monday, May 30, 2011

The Honorable Judge Wyoming
California Superior Court
San Francisco, Dept 15
850 Bryant Street
San Francisco, CA 94103

Re: REDUS, DORIAN
    SC#: *88778*

DOB: 05/19/46
CII: *M02858702*
PC 1026

*The Subject:* is necessary communication for the maintenance of my legal sanity.

*Legal Status:* I am a sixty-five-year-old African American male under a PC 187, murder. On August 9, 1974, I stabbed a woman I had known for six years to death. In October 1975, I was found not guilty by reason of insanity and remanded to Atascadero State Hospital. Then I was transferred to Napa State Hospital in 1982, and I was released on outpatient status in September 1988, under the auspices of San Francisco's CONREP. I believe it was in May of 2010 that you revoked my outpatient status.

Dear Judge Wyoming:

    I need to make one more very important truly stated statement of truth to terminate and end all my other truly stated statements of truth. It is my closing statement. Therefore, this whole four-part written plea, *A Quotidian Quash: From Mental Hygiene to Mental*

## A QUOTIDIAN QUASH: FROM MENTAL HYGIENE TO MENTAL HEALTH

*Health 1969–2011*, two previous letters, and this third and final letter regarding the commonplace situation below in a previous court where my defending attorney-at-law, Ira Barg, was provided to me by my San Francisco court, all of this is justly pleading to have all my sanity restored this year.

The following is a variation on the theme of my letter of Tuesday, January 16, 1996, to my former appeal attorney-at-law, Mark L. Christiansen:

> Although, I instigated no violence; what-so-ever, in my first fourteen years of hospitalization, I was involved, through no fault of my own, in the following below.
>
> *Napa State Hospital June 21, 1983*
>
> A tall incontinent (bed wetter) who slept in his bed next to my bed (every night) deposited such quantities of urine on the floor, under his bed, and wetting also his linen, I recall being the person who mopped the stench up every morning rather than smell it until noon when the janitors would begin to mop. [So, against my wishes, I was not anonymous to the incontinent patient.] Be that as it may, one day this tall male, from New York, attacked me [by the Ward Q 3 and Q 4 Office] as he thought I had his ball point pen. Which I did not. To cause us to fall as he yanked at the hair on my head, I hugged him and threw us both to the floor.
>
> *In Court on October 11, 1995*
>
> However, without defense October 11, 1995, when the district attorney-at-law for San Francisco, (named) Mr. Colorado, reiterated the one violent

scenario above, in his own closing statements, by the time he finished misquoting my testimony in his own twisted words—to my jury he had said, "You have heard Mr. Redus say he was so crazy, he attacked another patient because he thought the other patient's pen was a snake."

When Mr. Colorado was finished, I protested! I tapped my attorney, Ira H. Barg's wrist with the pads of my thumb and forefinger to secure from him, that he would object to the DA's lie! He said, "No!" Then when I asked my former attorney, Ira H. Barg, the attorney-at-law I had just before Cheryl H. Arkansas, to have my testimony read from that days court transcript he [again] said, "No!"

Instead of continuing with more travesties of justice like the above, please have my "court officers" start off by reading the following book. And as I am sane, please also provide me my United States constitutional guarantees—an immediate restoration of my sanity at the conclusion of a court hearing or a court trial. And please regard my finances too.

## A Good Book to Read during Deliberations

Primack, Joel R., and Nancy Ellen Abrams. *The View from the Center of the Universe: Discovering Our Extraordinary Place in the Cosmos*. New York: Riverhead Books, a Division of Penguin Group (USA), 2006.

I rest my forty-year-old case.

Respectfully submitted,

Mr. Dorian Gaylord Redus
Patient

Mr. Dorian Gaylord Redus
Ward T-15
Napa State Hospital
2100 Napa-Vallejo Hwy.
Napa, CA 94558-6234
1(707)252-9988

Monday, June 13, 2011

Dr. Eric Florida, MD
Ward T-15
Psychiatrist
Napa State Hospital
2100 Napa-Vallejo Hwy.
Napa, CA 94558-6234

Re: the need for a special conference on medications and rape.

Dear Dr. Florida:

    I gave Cory Alabama, CSW, a copy of my recent document, *A Quotidian Quash: From Mental Hygiene to Mental Health 1969–2011*, on Wednesday, May 11, 2011. All the letters cited in this letter are in the ward's copy of the document that I gave to Cory, where they are mostly listed chronologically.
    I would like to start with two key word idioms: (1) *in spite of*, *despite*, or *regardless of*, and (2) *because of*. There is a very short October 25, 2010, letter to Dr. Hameed Nebraska, MD, who was my Ward T-14 psychiatrist just before I came to Ward T-15 on September 15, 2010. The letter is about a short statement Dr. Hameed Nebraska made to me on Friday, October 22, 2010, as we passed on the grounds in front of Ward T-15. That was the day before Napa State Hospital staff person Donna Gross was murdered on the grounds on Saturday, October 23, 2010. Dr. Nebraska's statement was a suggestion, to me, that I take a "medication holiday" from my psychotropic medication. Currently and for some time now, I have been obviously doing better

and better in spite of, not because of, my psychotropic medication. I want you, my current prescribing physician, to consider me a candidate for a "psychotropic medication holiday."

I would like to officially hear from you in writing on my entire rape thirty years ago at California's Atascadero State Hospital in the early 1980s, and again at Napa State Hospital in the middle 1980s. They have never been discussed officially before me by any court officer. The rapes were not in spite of or regardless of my psychotropic medications. I maintain the stark rapes were because of my psychotropic medications, Prolixin from Dr. Wiggly at ASH and Haldol from Dr. Arizona at NSH. The psychology of the sex and a rudimentary synopsis of the psychiatry of the sex is described in my January 17, 2011, letter to the late Dr. Martin Luther King Jr. (written to him after his death as an attention-getter), and the description continues in the August 5, 1994, letter that follows the King letter in *A Quotidian Quash*.

When raping me with "big penises" was not enough, for no reason I know, see my Friday, January 7, 2011, letter to our own Dr. Mexico, PhD, and see also my Diatribe no. 7 of my Monday, November 15 letter to Cheryl H. Arkansas, you and CONREP have been aggressively raping me with "little penises" by investigating me for nonexistent child molestation for over five years. All I had to do was confess. No actual criminal act was necessary. You authorities would do the rest. It is a crying shame! Me, a child molester. You and CONREP are, in my angsty thoughts, both child molesters! It is easy to know why I was hearing angry voices from February 2009 after my sex offender treatment ended until I left CONREP for Napa State Hospital on October 1, 2009. Just read *A Quotidian Quash*.

In all of this, I see a (bad unilateral) threat of my comprehensive self-disclosure from me. However, in your office on Wednesday, June 1, 2011, you called for a team. I guess you need a special conference on medications and rape. I need something from you in writing. I have given Cory and you all I have to say in writing. Officially, please quid pro quo give me something in writing. When I see the word *rape* on page 8 of Mark Christiansen's appellant brief in *A Quotidian Quash*, I see the need for something in writing from my state hospital.

My life has become unmanageable in the past because of my psychotropic medications, and a psychotropic medication holiday from my psychotropic medication might help me to manage my life here and now on Napa State Hospital's Ward T-15.

I have said all of this to avert, not advert, our irreconcilable differences.

Thank you, and have a blessed day.

Respectfully submitted,

Mr. Dorian Gaylord Redus

Mr. Dorian Gaylord Redus
Ward T-15
Napa State Hospital
2100 Napa-Vallejo Hwy.
Napa, CA 94558-6234
1(707)252-9988

Sunday, July 17, 2011

To Whom It May Concern:

When I should be able to stop them, there are several things I have not stopped. I should be able to do something to stop them. However, all I can do is hurt. This hurting is foreign to me, and it has made me sick (mentally ill) in the past. It is verbal and physical rape, not therapy, that bothers me. And moreover, it bothers me that for some reason(s) unknown to me, my courts cannot handle the whole truth.

In my family, I often give a therapeutic shirt off my back, but I am more eased at that than I am eased at saying a shirt is not a shirt! Where ad hominem is an argument supported by adversarial prejudice and character assassination to avoid discussing the issues *(Dorian is delusional, and he writes bizarre letters)*. Where ad verecundiam is an argument that I will lose if I exhibit to much modesty *(I am delusional, and I write bizarre letters)*. And where ad baculum describes an inappropriate authority, because of my mental health system's past ad hominem and because of my past ad verecundiam, we are both an inappropriate authority! Yes, my adversaries have always used ad hominem to win in court, unless I used an attorney-at-law that I personally paid for being my personal attorney-at-law. When I have done that, I have always won! It should be simple to win again as in court, my adversaries tell lies (double entendre) above the law, and they use ad hominem. Yes, I have gone to court and listened to speculation, suspicions, conjecture, and adversarial lies take all my inalienable American rights. All of this malpractice is breaking my heart. I have spent over twenty years in state hospitals because my mental health system has made me sicker and sicker for over forty years.

I am not completely without fault. Although I never lost consciousness, I was once beat up through an army blanket at a military "blanket party" in the middle of the night, on a hill just outside of my Fort Polk Louisiana stateside barracks in 1967. The blows to my face and to my head through a thick army blanket by the fists of the solders, like the solders I was going to Vietnam with, left my face and my head all bruised up in the photographs that I took the very next day. They had all beat me at the blanket party to get me to confess to something that I had not even done, and it all made me angry.

*So it seems when I was tough in the face of tough US Army training for war and I did not confess to what I had not done, to Napa State Hospital's Dr. William Wisconsin, MD, decades later, it did not matter because I did not lose consciousness.*

> Mr. Redus has sustained no head injuries that whereby he lost consciousness... There is no family history of mental illness. The preceding two sentences are from page two of a January 25, 1995, letter to the Master Calendar Judge, San Francisco County Superior Court by Dr. William Wisconsin, MD.
>
> [Our tense and very stressful second wedding anniversary was on November 19, 1994.] On November 22, 1994 a Mini-Team Conference occurred to discuss the incident between he [Dorian Redus] and his wife, [Gillian Redus] which occurred on the unit during a visit. On this date Mr. Redus reported that he slapped his wife on her wrists... The preceding one and one half sentences are from page three of a January 25, 1995, letter to the Master Calendar Judge, San Francisco County Superior Court by Dr. William Wisconsin, MD.

When I was suffering some higgledy-piggledy (confusion) over my criminal (pun) justice system, the slap on my very frustrated wife's wrist was my accidental knee-jerk reaction to her forcefully taking two bottles from me during a visit when she had also missed my last court date because of her job. *By the time my tap-slap on Gillian's wrist was attested to in a San Francisco County Superior Court by Dr. William Wisconsin, MD, he had said, "Staff observed Mr. Redus beating his wife."*

The preponderating oxymoronic lies have been an overwhelming juggernaut that lies (double entendre) above our laws. I NEED AN ATTORNEY-AT-LAW! Even without an attorney-at-law, having sent most of these therapeutic letters and documents out according to the dates on them, none of the issues in *A Quotidian Quash: From Mental Hygiene to Mental Health 1969–2011* have been confirmed or denied or therapeutically replied to. Although what I want is involvement, so I may understand, not even one letter has been replied to in writing or otherwise.

Respectfully submitted,

Mr. Dorian Gaylord Redus
Patient

# FORENSIC HEALTH CARE INC.

March 22, 1994

Dorian Redus
6801 Mission Street #202
Daly City, CA 94014

Dorian:

I just read your letter. You say you didn't ever speak with Dr. Montana at our office. I fully believe you. I stand corrected. Please accept my apologies for this mis-statement. The next time we are in Court together have your attorney remind me and I'll correct the mistake for the record.

Sincerely,

Douglas R. Porky, PhD
Director
San Francisco/Marin Conditional Release Programs

DRK:st

---

Main Office: San Francisco Conditional Release Program, 110 Gough Street—Suite 301, San Francisco, California 94102 (415) 554-0305 FAX: (415) 554-0318
Satellite Office: Marln Conditional Release Program, P.O. Box 2728, San Rafael, California 94912 (415) 925-1380 FAX: (415) 925-1779

# FORENSIC HEALTH CARE INC.

Dorian Redus,

    As per your request, I am verifying that Dr. Douglas Porky received a telephone call on 2/23/94 from Dr. Montana. Dr. Montana related that he was very concerned that you, Dorian Redus, had attempted to contact him and had left him a message stating that you would be sending him, Dr. Montana, a parcel through the mail and needed his mailing address. Dr. Montana expressed concern and wanted to know what was going on with you, Dorian, and also informed Dr. Porky that he would be calling Arlo Smith, the DA.
    Dr. Porky in turn telephoned me and asked that I checkout why you were trying to get ahold of Dr. Montana and to inform you to desist from attempting to do so. I was able to reach you Friday 2/25/94 and relayed this message to you.

Janusz Mermel, LCSW

---

Main Office: San Francisco Conditional Release Program, 110 Gough Street—Suite 301, San Francisco, California 94102 (415) 554-0305 FAX: (415) 554-0318
Satellite Office: Marln Conditional Release Program, P.O. Box 2728, San Rafael, California 94912 (415) 925-1380 FAX: (415) 925-1779

Mr. Dorian Gaylord Redus
Ward T-15
Napa State Hospital
2100 Napa-Vallejo Hwy.
Napa, CA 94558-6234
1(707)252-9988

Sunday, July 17, 2011

## ASH First Day (circa) 1974

To Whom It May Concern:

    When I arrived at Atascadero State Hospital, I took a shower and changed on arrival. Then I went straight to the admission unit, immediately took medication (at the ward office), went to my room down a long hallway flanked by other rooms for other patients, and I lay down on my hospital bed in my new quarters to sleep. However, because that is not where I awoke, I reported where I awoke, and I asked for a much-needed explanation, which I have never gotten. Somehow, unknown to me, I had moved from my room halfway down the long hallway to the very back of the admission ward day room, and I awoke on a pile of mattresses. This story sometimes causes a curious perseveration on my first wakeup at California's Atascadero State Hospital.
    Thank you.

Respectfully submitted,

Mr. Dorian Gaylord Redus
Patient

Mr. Dorian Gaylord Redus
Ward T-15
Napa State Hospital
2100 Napa-Vallejo Hwy.
Napa, CA 94558-6234
1(707)252-9988

Sunday, July 17, 2011

## ASH Trust Office (circa) 1982

To Whom It May Concern:

    As I was starting to and preparing to slowly, carefully, and finally leave Atascadero State Hospital, I had an unforgettable problem with their African American trust officer. First, I got all the necessary permissions from whomever it concerned, then ordered some Dynaco (brand) electronic kits that needed assembly, having also arranged for a big assembly table in a big room I had easy access to. I ordered one preamp, one power amplifier, one FM tuner, and one electronic HiFi stereo equalizer. They were all 1980s-type high-end stereo system components that anyone might want. At hundreds of dollars each, it was quite a big deal for me and for my family. I had assembled Dynaco (home) electronics kits before. All I needed was a small soldering iron, a small wire cutter, some small screwdrivers, and about $800 to $1,000. It was all mail ordered. Second, when it had taken much too long to arrive, I was all the more patient. Then a small postcard came to me in the mail from an address near Atascadero State Hospital. It said they had all my Dynaco electronics kits, and they would delivered them to me at the hospital for an additional $54 extra shipping and handling charge. Third, I simply went to the hospital trust office where I had an account with more than sufficient funds. There, I asked the African American trust officer to remit the $54 amount so I could receive my stereo kits. He asked for the postcard. I simply handed him the postcard with all the information to receive my kits, and criminally, he did not give it back when I asked for it. It was a perfidious breach of trust because I never received my kits. *Fourth, when I shared all of this, face-to-face, with my father one evening at an Atascadero State Hospital "Black Project" family dinner, my father replied to me, "They are leaning on me."*

Thank you.

Respectfully submitted,

Mr. Dorian Gaylord Redus
Patient

All I need is my court-ordered sanity restored, and my court-ordered full compensation for all the malpractice. The first copying of this whole document, *A Quotidian Quash: From Mental Hygiene to Mental Health 1969–2011*, went to the copier on April 12, 2011, and this second copying including the documents in part 2 will go to the copier about three or four months thereafter.

## A QUOTIDIAN QUASH: FROM MENTAL HYGIENE TO MENTAL HEALTH

State of California—Health and Human Services Agency
FS002 (04/93)

Department of Developmental Services
Fax number (916) 653-4587

### STATEMENT OF ACCOUNT

STATEMENT DATE: 07/08/11

| CLIENT'S NAME | HOSP NO. | ACCOUNT NUMBER | PAYOR NO. | PERIOD COVERED BY THIS STATEMENT | |
|---|---|---|---|---|---|
| REDUS DORIAN G | 04 | 04208235-9274 | 110 | 06/01/11 | 06/30/11 |

INDICATE CLIENT'S NAME, HOSPITAL NUMBER, ACCOUNT NUMBER AND PAYOR NUMBER ON ALL INQUIRIES AND PAYMENTS.

TRANSACTIONS MADE AFTER THIS DATE WILL APPEAR ON YOUR NEXT STATEMENT

PAY THIS AMOUNT: 49,833.00

CHECK THIS BOX IF PAYING THE FULL AMOUNT STATED IN THE "PAY THIS AMOUNT" BOX ☐

**BILL TO**
DORIAN REDUS
C/O NAPA STATE HOSPITAL
2100 NAPA-VALLEJO HWY
NAPA                CA    94558

**REMIT TO**
DEPARTMENT OF DEVELOPMENTAL SERVICES
Accounting Section
P. O. Box 944202
Sacramento, CA 94244-2020

Make checks payable to:
DEPARTMENT OF DEVELOPMENTAL SERVICES

IMPORTANT: DETACH AND RETURN THE TOP PORTION OF THIS STATEMENT WITH YOUR REMITTANCE TO ASSURE PROPER CREDIT

| CLIENT'S NAME | HOSP NO. | ACCOUNT NUMBER | PAYOR NO. | ACTUAL COST BALANCE | STATEMENT DATE | PAGE NO |
|---|---|---|---|---|---|---|
| REDUS DORIAN G | 04 | 04208235-9274 | 110 | 415,527.42 | 07/08/11 | 1 |

| TRANSACTION DATE | REFERENCE NUMBER | DESCRIPTION | INSURANCE PORTION | AMOUNT |
|---|---|---|---|---|
| 07/07/11 | | MONTHLY AMOUNT | | 2,373.00 |

PAYMENT IS DUE IMMEDIATELY UPON RECEIPT OF THIS STATEMENT.

IF YOU HAVE ANY QUESTIONS CONCERNING THIS STATEMENT, PLEASE CONTACT
PHONE:

INFORMATION REGARDING THIS STATEMENT, ACCOUNT CHARGES, BALANCE, OR ANY OTHER MATTER MUST BE DIRECTED TO THE TRUST OFFICE AT 707-254-2424. THE DEPARTMENT OF DEVELOPMENTAL SERVICES IS UNABLE TO ASSIST YOU WITH ANY INFORMATION REGARDING YOUR ACCOUNT.

| ACCOUNT BALANCE LAST STATEMENT | 47,460.00 |
|---|---|
| NEW CHARGES/ADJUSTMENTS | 2,373.00 |
| NEW PAYMENTS/CREDITS | 0.00 |
| CURRENT ACCOUNT BALANCE | 49,833.00 |

REDUS DORIAN G        04 04208235-9274 110

PAY THIS AMOUNT: 49,833.00

IMPORTANT INFORMATION ON REVERSE SIDE

CONFIDENTIAL CLIENT INFORMATION
SEE CALIFORNIA WELFARE AND INSTITUTIONS CODE
SECTIONS 4514 OR 5328

# THE LAST WORD

My whole de facto (cost of care) debt is c. $1,000,000, and what little of that amount that I have is going for my attorney-at-law's fees. So my side, my ex-wife's ex-husband, my nuclear family, and my extended family and my friends won't have an unwanted putative queer sexual pervert, me, to relate to.

# PART THREE

Mr. Dorian Gaylord Redus
Ward T-15
Napa State Hospital
2100 Napa-Vallejo Hwy.
Napa, CA 94558-6234
1(707)252-9988

Monday, June 20, 2011

Re: us agreeably coming to accord on bin Laden's fate and accepting tutelage.

Dear Brother:

I found your address and phone number in an old letter of Tuesday, June 16, 1998. It was from me to you saying I would never call you or write you again. I trust you as I have changed. I am sane, and this year, I need a therapeutic attorney-at-law.

What do you do every day? Teach? Practice law?

It is Saturday, June 11, 2011. There was a hospital-wide family barbeque from 9:30 a.m. to 1:30 p.m. today. I did not go. However, I did go to the store. It is called the By Choice Store. I found a favorite issue of *National Geographic*, July 2010, for fifty points. I get one hundred points a day (free) for full participation in the therapy here. The magazine is on the evolutionary road to us from a 4.4-million-year-old fossilized woman, Ardi, an *Ardipithecus ramidus*, discovered by the University of California at Berkeley's Tim White and his Ethiopian colleagues. Ardi was discovered circa 1994. She is older than Lucy, an *Australopithecus afarensis* discovered by Donald Johanson in 1974. Lucy is 3.2 million years old.

> THE HUMAN FAMILY: The record of our lineage in Africa now extends back over six million years. The Middle Awash of Ethiopia has yielded fossils of all three major phases in hominid evo-

lution—Ardipithecus, Australopithecus, and Homo. (Page 45)

It is Sunday, June 12, 2011. Don't wear brown or khaki, and bring a driver's license if you want to visit me here anytime soon. The visiting is only on Thursday, Friday, Saturday, and Sunday from 9:00 a.m. to 5:30 p.m. You can see my cleanly shaved head. In my family, I often give a therapeutic shirt off my back, but I am more eased at that than I am eased at saying a shirt is not a shirt! Where ad hominem is an argument supported by adversarial prejudice and character assassination to avoid discussing the issues *(Dorian is delusional, and he writes bizarre letters)*. Where ad verecundiam is an argument that I will lose if I exhibit too much modesty *(I am delusional, and I write bizarre letters.)* And where ad baculum describes an inappropriate authority, because of my mental health system's past ad hominem and because of my past ad verecundiam, we are both an inappropriate authority! Yes, my adversaries have always used ad hominem to win in court, unless I used an attorney-at-law that I personally paid for being my personal attorney-at-law. When I have done that, I have always won! It should be simple to win again as in court my adversaries tell lies (double entendre) above the law, and they use ad hominem. Yes, I have gone to court and listened to speculation, suspicions, conjecture, and adversarial lies take all my inalienable American rights. All of this malpractice is breaking my heart. I have spent over twenty years in state hospitals because my mental health system has made me sicker and sicker for over forty years.

## A Controversial Action and Accomplishment

I ask where the elimination of the American enemy, Osama bin Laden, compliments the United States? Were two US presidents, first George Bush and second Barack Obama, appropriate to go after bin Laden? I also ask, is the whole topic too bellicose, too much revenge taking, and too warlike to even discuss? Our democratic president Barack Obama's cup "runs over," and some people even think former Republican President George Bush, who tried to get the American enemy Osama bin Laden and failed, should be considered in on the successful killing of the American enemy, Osama bin Laden. This subject matter is belligerent, but is it a killing most of America, and many cosmopolitan peoples of the world, may agree on? I think in Latin, *compere:* fill up the world's cup with joy and rejoice! The one responsible for "masterminding" 911 is gone! I also say my compliments, good wishes, and regards to our elite Navy SEAL commandos who did the dirty deed!

> At last, bin Laden is gone
> Osama bin Laden: The end of a 10-year manhunt

What happened

> After a decade on the run, Osama bin Laden was killed this week in a daring U.S. special operations forces raid on his hideout in a custom-built, walled compound in Pakistan. Acting on orders from President Obama, 79 elite Navy SEALs descended by helicopter on Abbottabad, an affluent suburb 35 miles north of the Pakistani capital of Islamabad, which U.S. intelligence had pinpointed as the likely hiding place of the al Qaida founder and leader. Following a 40-minute firefight—during which U.S. officials said five people were killed, including bin Laden's son Hamza—bin Laden was found in an upstairs bedroom. He was unarmed but attempted to resist capture, U.S. officials said, and was killed by a shot to the head. The SEALs left the compound with a trove of computer hard drives, thumb drives, and documents, as well as bin Laden's corpse. His body was later buried in the Arabian Sea by the U.S. aircraft carrier *Carl Vinson,* so as to deny his followers a shrine. (Page 4 of the May 13, 2011, volume 11, issue 514 of *The Week,* a weekly news media magazine)

*We got him'*
*How painstaking intelligence work—and high-stakes decisions—ended the hunt for Osama bin Laden*

For years, THE agonizing search for Osama bin Laden kept coming up empty. Then last July, Pakistanis working for the Central Intelligence Agency drove up behind a white Suzuke navigating the bustling streets near Peshawar, Pakistan, and wrote down the car's license plate. The man in the car was bin Laden's most trusted courier, and over the next month CIA operatives would track him throughout central Pakistan. Ultimately, he led them to a sprawling compound at the end of a long dirt road and surrounded by tall security fences in a wealthy hamlet 35 miles from the Pakistani capital.

On a moonless night eight months later, 79 American commandos in four helicopters descended on the compound, officials said. Shots rang out. A helicopter stalled and would not take off. Pakistani authorities, kept in the dark by their allies in Washington, scrambled forces as the American commandos rushed to finish their mission and leave before a confrontation. Of the five dead, one was a tall, bearded man with a bloodied face and a bullet in his head. A member of the Navy SEALs snapped a picture with a camera and uploaded it to analysts who fed it into a facial-recognition program.

And just like that, history's most expansive, expensive, and exasperating manhunt was over.

"We have a visual on Geronimo [Osama bin Laden],"... A few minutes later: "Geronimo

EKIA." Enemy Killed in Action. There was silence in the Situation Room.

Finally, the president spoke up.

"We got him." (Page 44 of the May 13, 2011, volume 11, issue 514 of *The Week*, a weekly news media magazine)

May our spiritual love make us one in our support of America's deed! Have a blessed day.

Respectfully submitted,

Mr. Dorian Gaylord Redus
Your little brother

Mr. Dorian Gaylord Redus
Napa State Hospital
Ward T-15
2100 Napa-Vallejo Hwy.
Napa, CA 94558-6234
1(707)252-9988

Friday, July 15, 2011

M. Mary Oregon
Director Mall Program Services
Napa State Hospital
2100 Napa-Vallejo Hwy.
Napa, CA 94558-6234

To Whom It May Concern:

    Thank you for (recently) helping me to get my chalk abstract artwork from Arts in Mental Health. It now has the most prominent position in my four-man dormitory. It is also most immediately noticeable and conspicuous to all my other (wall) mounted artwork. I call it *Abstract and Probe*, after NASA's 2001 WMAP satellite space probe.
    Much more than thanking you, I have a simple request. I have done all my printing at the S8 Computer Lab lately during mall hours. I have actually been working for months on a 130-some-page very helpful manuscript. It is almost finished. Plus, I plan to get ten bound copies of it. Where *A Quotidian Quash: From Mental Hygiene to Mental Health 1969–2011*, its title, translates to a commonplace silencing, put down, annulment, or suppressing. Please peruse all of my manuscript, *A Quotidian Quash: From Mental Hygiene to Mental Health 1969–2011*, even though it is just 130 pages of letters that tell my therapeutic story. And then let me know all of your feelings in writing. It is a really big and abstruse request until you have done all the reading.
    I seek the whole truth through you in court. Please send me a letter for my court documenting my "no problem," very easy to

please, and exemplary behavior: printing. Just ask the mall assistant, Imelda, who often escorted me and also occasionally supervised my printing at the S8 (Crossroads) Computer Lab, or ask her and Jackie Finch when she returns from her vacation.

When I get the ten bound copies, if I don't hear from you first, then I will still get one copy to you. I very much want your esteemed thoughts and feelings on what I have done under your direction.

Thank you again.

Respectfully submitted,

Mr. Dorian Gaylord Redus
Patient

Disability Programs
and Resource Center
1600 Holloway Avenue,
Student Services Building 110
San Francisco, CA 94132
Tel: 415/338-2472

Dorian Redus
Ward T-15
Napa State Hospital
2100 Napa-Vallejo Hwy.
Napa, CA 94558-6234

July 21, 2011

Dear Dorian,

I received and read your March 28 memo addressed to me, and was not able to find the faculty that you mentioned to forward your manuscript to. I am sorry that I can't be more help in this matter.

Sincerely,

Deidre Defreese

Mr. Dorian Gaylord Redus
Napa State Hospital
Ward T-15
2100 Napa-Vallejo Hwy.
Napa, CA 94558-6234
1(707)252-9988

Monday, August 1, 2011

Medical Records
Napa State Hospital
2100 Napa-Vallejo Hwy.
Napa, CA 94558-6234

To Whom It May Concern:

When I was on our old C-ward in *1988*, I had a psychologist's *evaluation*. It was a test.

I would like to have and to read my IQ score and my test's report by Dr. Kepner, PhD, who still works here at Napa State Hospital as the staff psychologist on our (current) Ward T-16.

Before the test in 1988, Dr. Kepner, PhD, said I would be given my test's results, and he said I would be followed. However, to this day, every time I ask Dr. Kepner for the results and my test score, which I have never received, he makes a balk (like in baseball) and only says, "You are a smart guy." And he then simply walks away smilingly.

Please have the doctor or Medical Records send me the score and the report. As I recall, the intelligence test I took was a *Wechsler Adult Intelligence Scale*, which may be converted to a simple IQ score.

I need the insight the (promised) IQ score will provide me, just like Dr. Kepner did, so I do not draw any prejudiced conclusions.

I understand I will be charged ten cents per page for my copies. Thank you.

Respectfully submitted,

Mr. Dorian Gaylord Redus
Patient

Cc: Virginia Illinois
    Social Worker, Ward T-15

# A QUOTIDIAN QUASH: FROM MENTAL HYGIENE TO MENTAL HEALTH

Mr. Dorian Gaylord Redus
Ward T-15
Napa State Hospital
2100 Napa-Vallejo Hwy.
Napa, CA 94558-6234
1(707)252-9988

Friday, August 12, 2011

Dr. Eric Florida, MD
Ward T-15
Psychiatrist
Napa State Hospital
2100 Napa-Vallejo Hwy.
Napa, CA 94558-6234

Re: the general request I made on Wednesday, August 10, 2011, being referred to my team. Please find attached one detailed (letter type) order form.

Dear Dr. Florida:

    The documents I asked for in my (recent) general request to you are needed. My document is needed in multiple copies so all court officers on my side, the defense, may see and attest to my sanity at my next hearing or trial, hopefully this year or as soon as possible. At some previous hearings and trials, adversarial court officers for the prosecution have described my writing as bizarre and described my main points as delusional. The manuscript I am requesting at my personal expense from our local FedEx office copier and binder will stop the PC 1370 type (unable to stand trial) exacerbation on both sides of my judge's bench which has, in the past, worked against me because I could not communicate my main points without my new manuscript. When I and my court officers read my new manuscript as a call to action, it will guarantee my sanity at my next court hearing or trial.

I can show you an excellent example of FedEx Office's excellent work.

May I please order and receive my documents at $250?

Respectfully submitted,

Mr. Dorian Gaylord Redus
Patient

Mr. Dorian Gaylord Redus
Napa State Hospital
Ward T-15
2100 Napa-Vallejo Hwy.
Napa, CA 94558-6234
1(707)252-9988

Tuesday, August 16, 2011

Dr. Carol Carolinas, PhD
Napa State Hospital
Psychologist
Ward T-14
2100 Napa-Vallejo Hwy.
Napa, CA 94558-6234

To Whom It May Concern:

Thank you for receiving this therapeutic manuscript regarding murder, medications, rape, and a really ironic, unjust, and oxymoronic kangaroo court.

I have a super plan for getting myself out of Napa State Hospital for good. This manuscript, *A Quotidian Quash: From Mental Hygiene to Mental Health 1969–2011*, is the cornerstone of my sanity. And the new inserted document, *A Three-Part Discussion*, is the cornerstone of *A Quotidian Quash*. The middle section of *A Three-Part Discussion*, RCTVU (relativistic color television), is the cornerstone of *A Three-Part Discussion*. The paragraph on the lower half of RCTVU's page 7 is the therapeutic cornerstone of the whole RCTVU section. And furthermore, note 2 in *A Three-Part Discussion* is the cornerstone of the paragraph on the lower half of RCTVU's page 7. I know it is a long chain of cornerstones!

However, you now have the (therapeutic) first three quarters of a document for my court officers. When I get an attorney, my attorney's doctors may read their 132-page copies of my new manuscript, *A Quotidian Quash: From Mental Hygiene to Mental Health*

*1969–2011*, hopefully understand my sanity, and publicly attest to it under oath.

Thank you.

Respectfully submitted,

Mr. Dorian Gaylord Redus
Patient

# A QUOTIDIAN QUASH: FROM MENTAL HYGIENE TO MENTAL HEALTH

Mr. Dorian Gaylord Redus
Napa State Hospital
Ward T-15
2100 Napa-Vallejo Hwy.
Napa, CA 94558-6234
1(707)252-9988

Friday, August 19, 2011

Virginia Illinois
Social Worker
Napa State Hospital
Ward T-15
2100 Napa-Vallejo Hwy.
Napa, CA 94558-6234

U*RGENT*

To Whom It May Concern:

Since the hospital-wide family barbecue is on Saturday, September 10, 2011, I am very fast running out of time to make the personal arrangements for my daughter and my two granddaughters to attend and support me in my efforts to live in this state hospital, enjoy the summer cuisine and repast, and at the same time mentally keep up with my daily responsibilities as a father and grandfather of two. Even though, Susanna, I have been bugging you regarding the barbecue for weeks, we still need to get our gears in motion before it is too late.

Today, please do get me the form that I need. It usually tells the amount necessary for each adult to pay for the nice meal, requests driver's license numbers for all attending adults, and also requests the address and the phone numbers of all attending from the community. Please also do note that my youngest granddaughter is over twenty-one years of age.

Thank you.

Respectfully submitted,

Mr. Dorian Gaylord Redus
Patient

Mr. Dorian Gaylord Redus
Napa State Hospital
Ward T-15
2100 Napa-Vallejo Hwy.
Napa, CA 94558-6234
1(707)252-9988

Friday, August 19, 2011

Michael Finney
7 On Your Side
ABC7 Broadcast Center
900 Front Street
San Francisco, CA 94111
1(415)954-7777

Dear Mr. Michael Finney:

    Thank you for any help that you were in my getting my old (mental health) IQ scores of 107, 109, and 113 from 1988 and 1985 (bright average). Having them is helpful. Having just returned to Napa State Hospital on October 1, 2009, on October 7, 2009, my results were "High Average—Superior range of intelligence."
    My current Ward T-15 psychiatrist and social worker putatively helped me to get all the test results above. Furthermore, if you helped too, then again, thank you. Kudos for 7 On Your Side!
    I received the two requested reports from the 1980s on Wednesday, August 17, 2011. They were so helpful to me. I am also trying to get my October 7, 2009, results and its report by Napa State Hospital's Dr. Dakota.
    I have prevailed against intransigent situational injustice.

Respectfully submitted,

Mr. Dorian Gaylord Redus
Patient

Mr. Dorian Gaylord Redus
Napa State Hospital
Ward T-15
2100 Napa-Vallejo Hwy.
Napa, CA 94558-6234
1(707)252-9988

Thursday, September 1, 2011

Dr. Carol Carolinas, PhD
Napa State Hospital
Psychologist
Ward T-14
2100 Napa-Vallejo Hwy.
Napa, CA 94558-6234

To Whom It May Concern:

    Thank you for receiving my very important one-hundred-page spiral-bound manuscript regarding murder, medications, rape, and a really ironic, unjust, and oxymoronic kangaroo court. I sent it to you on Tuesday, August 16, 2011, with a strong request that you acknowledge its receipt. So please do acknowledge your receipt of my valuable manuscript. And again, thank you for receiving it.

    You said before that you did not get my other fifteen-page treatise for months, as it was in the wrong box, and I write you now because there may be another problem with your mail service.

    Thank you.

Respectfully submitted,

Mr. Dorian Gaylord Redus
Patient

Mr. Dorian Gaylord Redus
Napa State Hospital
Ward T-15
2100 Napa-Vallejo Hwy.
Napa, CA 94558-6234
1(707)252-9988

Thursday, September 22, 2011

The San Francisco Chronicle
901 Mission Street
San Francisco, CA 94103-2988
1(415)777-1111

To Whom It May Concern:

    I want the name of a person at the *San Francisco Chronicle* address above to send my written work to. I have a photocopied and spiral-bound manuscript of 132 pages in length. *My manuscript may have some news value, editorial value, or my manuscript may be so useful to you that you will want to advise its publication in its entirety.* For now, I just need the name of a nice person at the *San Francisco Chronicle* address above that I may send my manuscript to.
    I feel that someone currently powerful and decisive at the *San Francisco Chronicle* should experience firsthand my manuscript's firsthand descriptions of my life here as a patient from 1982 to 2011 by carefully reading all the new manuscript that I want to send to you for free. My August 9, 1974, PC 187, murder, which was a front-page news story in the *Chronicle*, is described in my manuscript. My 1983 homosexual raping due to Napa's mandatory abject psychotropic drugs is described in my manuscript. And the nontherapeutic injustice (from those lying double entendre above the law) that has preponderated at my many San Francisco Superior Court's hearings and trials is also described very well in my manuscript.

Please forward a name to me as soon as possible.
Thank you.

Respectfully submitted,

Mr. Dorian Gaylord Redus
Patient

Mr. Dorian Gaylord Redus
Napa State Hospital
Ward T-15
2100 Napa-Vallejo Hwy.
Napa, CA 94558-6234
1(707)252-9988

Thursday, September 22, 2011

M. Mary Oregon, PhD
Director Mall Program Services
Napa State Hospital
2100 Napa-Vallejo Hwy.
Napa, CA 94558-6234

Re: REDUS, DORIAN					DOB: 05/19/46
    SC#: 88778					PC 1026

*Legal Status:* I am a sixty-five-year-old African American male under a PC 187, murder. On August 9, 1974, I stabbed a woman I had known for six years to death. In October 1975, I was found not guilty by reason of insanity and remanded to Atascadero State Hospital. Then I was transferred to Napa State Hospital in 1982, and I was released on outpatient status in September 1988, under the supervision of San Francisco's CONREP. On October 1, 2009, I returned to NSH for keeping the angry voices I was hearing for months a secret. And on May 5, 2010, my outpatient status was revoked. Now I want to win a sanity hearing or trail this year, 2011.

To Whom It May Concern:

    Thank you for receiving this therapeutic manuscript regarding murder, medications, rape, and some really ironically unjust, oxymoronic kangaroo courts.
    This unlikely and wonderful chance to give you this 130-page manuscript is a "quantum leap" for Napa State Hospital. As you

read every page of this manuscript, please do not treat me as if the only thing I am good for is being an example of "good for nothing," because that will, very regrettably, make you, in my court affairs, "good for nothing." Don't forget this is all where my *A Quotidian Quash* translates into a commonplace silencing.

And please do not forget I very much want your esteemed thoughts and feelings on what I have done here on computer printers under your tutelage and direction.

Thank you again.

Respectfully submitted,

Mr. Dorian Gaylord Redus

Mr. Dorian Gaylord Redus
Napa State Hospital
Ward T-15
2100 Napa-Vallejo Hwy.
Napa, CA 94558-6234
1(707)252-9988

Monday, September 26, 2011

Please find enclosed one (132-page) manuscript.

Re: REDUS, DORIAN            DOB: 05/19/46
    SC#: 88778                CII: M02858702
                              PC 1026

*The Subject:* is a request that you please find me a winning attorney-at-law.

*Legal Status:* I am a sixty-five-year-old African American male under a PC 187, murder. On August 9, 1974, I stabbed a woman I had known for six years to death. In October 1975, I was found not guilty by reason of insanity and remanded to Atascadero State Hospital. Then I was transferred to Napa State Hospital in 1982, and I was released on outpatient status in September 1988, under the supervision of San Francisco's CONREP. On October 1, 2009, I returned to NSH for keeping the angry voices I was hearing for months a secret. And on May 5, 2010, my outpatient status was revoked. Now I want to win a sanity hearing or trail this year, 2011.

Dear Friend:

When I have actually been the only sane American in my courtrooms without this new manuscript, those that hear me speak in my next courtroom will think my speaking out evidences my usual crazy verbiage, and the court will, no doubt, find that I am insane. And therefore, my court will conclude to retain and treat. However, to begin my defense and win a SANITY RESTORED VERDICT, thank God I have lived and finished this manuscript.

Please read all of this letter, and please also have the attorney that you are referring read it and also read all of my 130-page manuscript before they take my case. This should be an "open and shut" pro bono affair, but I can probably, if necessary, come up with (my life savings) a few tens of thousands of dollars for some expert witnessing psychiatric testimony, attorney-at-law fees, and perhaps some other stellar witnesses.

With all my previous court-paid public defenders, the preponderating oxymoronic lies have been an overwhelming juggernaut that lies (double entendre) above our laws. I NEED AN ATTORNEY-AT-LAW THAT I PAY! Even without a paid attorney-at-law, I sent out most of the therapeutic letters and the therapeutic documents according to the dates on them, yet none of the issues in my manuscript, *A Quotidian Quash: From Mental Hygiene to Mental Health 1969–2011*, have been confirmed or denied or even therapeutically replied to in writing. What I want is involvement, and I need an end to the injustice in my courtrooms.

I need to make one more very important truly stated statement of truth to terminate and end all my other truly stated statements of truth in my manuscript. It is my closing statement. George Carlin is quoted in the July 1–8, 2011, issue of *The Week* media magazine as saying: "Some people see things that are and ask, Why? Some people dream of things that never were and ask, Why not? Some people have to go to work and don't have time for all that." Witnessing without even the internet, I seem to have made a slight mistake (a picayune peccadillo) in my manuscript, where I wrote: "Supposedly, former President John Fitzgerald Kennedy said, something like 'Some people see things that are, and they ask, why? I see things that never were, and I ask why not?'" Moreover, Marshall McLuhan is quoted in the April 22, 2011, issue of *The Week* as having said, "There are no passengers on Spaceship Earth. We are all crew." I wrote honestly from deep in my own heart instead of accurately quoting former President John Fitzgerald Kennedy.

I feel/think this polite member of the crew has respectfully submitted this fantastic winning California notion, my pipe dream, for a winning attorney-at-law, hoping that it is not an empty, false hope pie in the sky.

Without the internet, I say politely my environment is queer. Or better, I say my environment begins when I get a winning attorney-at-law. I hope that you may find me an adequate attorney-at-law for my special needs, like my high average-superior range of intelligence and my average funding. In final conclusion, thank you very much for any help in my personal fight against the injustice in my San Francisco courtrooms. Maybe your friend Mariannie may be of service in this necessary endeavor.

Thank you.

Respectfully submitted,

Mr. Dorian Gaylord Redus
Patient

EDITED COPY

Mr. Dorian Gaylord Redus
Napa State Hospital
Ward T-15
2100 Napa-Vallejo Hwy.
Napa, CA 94558-6234
1(707)252-9988

Wednesday, September 28, 2011

Dr. Eric Florida, MD
Ward T-15
Psychiatrist
Napa State Hospital
2100 Napa-Vallejo Hwy.
Napa, CA 94558-6234

Re: more intelligent and continuing communication for the maintenance of my sanity. Please find attached one 132-page document to help us to prepare for court.

Dear Dr. Florida:

Being nebulous and also speaking generally about many specific verbal and written reports and speaking also of the individual, thinking, of my past psychiatrists: their nontherapeutic putative focuses or foci have been punitive.

Dr. G. wrote: "He [Dorian] was found Not Guilty by Reason of Insanity on October 29, 1974." Then in the next paragraph on the same page, Dr. G. wrote: "In October of 1975, he [Dorian] was determined to be Not Guilty by Reason of Insanity." These two one-sentence quotes are from page 1 of a January 25, 1995, letter to the Master Calendar Judge, San Francisco County Superior Court by Dr. William Wisconsin, MD, formerly a staff psychiatrist of Napa State Hospital. There was no recidivism. The two different dates are for the same NGRI judgment. Thus, they are an erratic erratum. Furthermore, on page 3 of that January 25, 1995, letter to the Master

Calendar Judge, San Francisco County Superior Court, Dr. G. also wrote: "Mr. Redus [Dorian] is alert and orientated but a poor historian." My point here is that Napa State Hospital's Dr. G. should have written: "Mr. Redus is alert, orientated, and intelligent."

If I stand alone in a San Francisco Superior Court without you and my manuscript, *A Quotidian Quash*, then Dr. William Wisconsin's words—"poor and historian"—are an interesting putative (choice) that is a punitive (choice of words) to punish me and to put me down. If I personally purport, posit, and behest that he should have more correctly used the words *poor and historian* to modify himself, then without you and my manuscript, *A Quotidian Quash*, what I say here and in a San Francisco Superior Court "puts the shoes of the wrong doer and mental illness on me in an ironic treacherous travesty of justice." This is the same conjecture that I made in my Wednesday, October 20, 2010, letter to you, which is in my manuscript.

*A Quotidian Quash: From Mental Hygiene to Mental Health 1969–2011* has some serious sociology, psychology, and psychiatry that effectively changes the scales of justice in many of my past legal proceedings. And therefore, it is about a forensic debate, even though it is just a therapeutic treatise for the maintenance of my personal sanity in my hands. In your more respectfully regarded hands, you could use it at a San Francisco County Superior Court to more respectfully regard me and make me a lot of cash money. And through you and other respectfully regarded witness participants there, where I want to be with you, you and they could testify for me, such that I may win a nice sanity restored verdict. If you appropriately help me to use my manuscript by carefully reading all of it, I will never be the unknown and daft quotidian dogface veteran that I am now. There, where I want to be with you, I will be an intelligent long-term psychiatric patient.

What happens if you or I think you or I are trying to cause trouble? That is what my manuscript changes in our lives. Where you and your California state hospitals and mental health system claim that I owe you/them $1 million, my manuscript levels the playing field.

It is one step forward and three steps backward. You (and your guys) owe me at least $2 million.

> You walk up to the gate into yesterday, and as you approach you see a version of yourself waiting for you there, looking about one day older than you presently are. Since you know about the closed time like curves, you are not too surprised; obviously you lingered around after passing through the gate, looking forward to the opportunity to shake hands with a previous version of yourself. So the two versions of you exchange pleasantries, and then you leave your other self behind as you walk through the front of the gate into yesterday. But after passing through, out of sheer perverseness, you decide not to go along with the program. Rather than hanging around to meet up with your younger self, you wander off, catching a taxi to the airport and hopping on a flight to the Bahamas. You never do meet up with the version of yourself that went through the gate in the first place. But that version of yourself did meet with a future version of itself—indeed, you still carry the memory of the meeting. What is going on?
>
> ONE SIMPLE RULE
> There is a simple rule that revolves all possible time travel paradoxes. Here it is:
>
> - Paradoxes do not happen.
>
> It doesn't get much simpler than that. (Sean Carroll, *From Eternity to Here: The Quest for the Ultimate Theory of Time*)

Innocent: I would rather (intellectually) involve myself in the following, from above, rather than involving myself in my unjust frame-up on your trumped-up charges of child molestation for another five years.

> Rather than hanging around to meet up with your younger self, you wander off, catching a taxi to the airport and hopping on a flight to the Bahamas. You never do meet up with the [older] version of yourself that went through the gate in the first place. But that [older] version of yourself did meet with a future [older] version of itself—indeed, you still carry the memory of meeting. What is going on?

If the police outside a hypothetical gate to yesterday (crossroad) at this complex did not have me here and communicating, then the police would have to come through a "closed time like curve" at a hypothetical gate to yesterday to arrest you and others here for jailing, *in saecula saeculorum*, for ever and ever, without your full free will. My check and my balance is all that is here communicating and adversarial to you here. By my articulate manuscript, I am not a "poor historian." I am articulate. Even the Master Calendar Judge at a San Francisco Superior Court and Dr. William Wisconsin, MD, formerly a staff psychiatrist at Napa State Hospital, may be made to understand that I am not a "poor historian." And if you carefully read all of my 132-page manuscript, *A Quotidian Quash*, you, too, may be convinced of the errors in our Napa State Hospital's ways and conclude that my sanity is restored.

In it, my readers may see prejudice and character assassination in issues like *Dorian is delusional, and he writes bizarre letters* and *I am delusional, and I write bizarre letters*. Where ad baculum is an inappropriate authority that in the past has intimately maintained that my two therapeutic theories STS (space-time sphere) and RCTVU (relativistic color television universe) are (de facto) my well-developed delusional theories on my autistic universe, you should read my very appropriate written work and then conclude otherwise for me in court. As my prescribing and diagnosing psychiatrist for over one whole year now, you have ignored my written requests that you address my homosexual rape issues due to abject psychiatry in writing. You should change AND read my very appropriate work, and you should also comment to me in writing. Like I am not a "poor historian," I am not in any way a very clinically delusional, bizarre, or in any way evil patient of yours, as you will see if you just read the attached document clear cover to clear cover to help us to both prepare intelligently for my next court as soon as possible.

Generally, my behaviors and my thoughts fascinate me to no end, and I am a happy fellow surrounded by other happy male and happy female fellows that include happy staff when I am here. Please just make sure I receive a nice copy of my Ward T-15 Team's next annual letter to the Master Calendar Judge, San Francisco County Superior Court regarding my PC 1026, murder. It will be a compromise that will restart, and perhaps it will be sufficient.

Thank you.

Respectfully submitted,

Mr. Dorian Gaylord Redus
Psychiatric patient

Cc: Dr. Mexico
　　Psychologist, Ward T-15

Cc: Virginia Illinois
　　Social Worker, Ward T-15

Mr. Dorian Gaylord Redus
Napa State Hospital
Ward T-15
2100 Napa-Vallejo Hwy.
Napa, CA 94558-6234
1(707)252-9988

Monday, October 3, 2011

Virginia Illinois
Social Worker
Napa State Hospital
Ward T-15
2100 Napa-Vallejo Hwy.
Napa, CA 94558-6234

To Whom It May Concern:

One of my reasons for not paying all my "cost of care" bills is the following five harmful policies of this California state hospital for the criminally and mentally insane.

1. I am not criminally or mentally insane, yet I am here. Yes, I have a mental illness. I have given my psychiatrists too much information, but I am not afflicted with any of their suffering (because) I am mostly in remission, even though my psychiatrists have not owned many of their harmful treatments in courtrooms where I have been.
2. This state hospital has a twenty-some-year-old policy of a dangerous denial of having homosexually raped me with its abject psychotropic medication in c. 1983.
3. Napa has a very disabling policy about my two visionary scientific works: RCTVU (relativistic color television universe) and STS (space-time sphere) theory. Napa just ignores them.

4. Napa has a putative history of participation in "kangaroo courts;" and therefore the preponderances in my courtrooms lie (double entendre) above the law.
5. For some twenty plus years, this state hospital denied me the results it promised me on my 1988 *Wechsler Adult Intelligence Scale*, and at this crossroad in my life, I believe that my IQ is helpful to me.

Thank you.

Respectfully submitted,

Mr. Dorian Gaylord Redus
Psychiatric patient

Mr. Dorian Gaylord Redus
Napa State Hospital
Ward T-15
2100 Napa-Vallejo Hwy.
Napa, CA 94558-6234
1(707)252-9988

Tuesday, October 4, 2011

Re: *A Quotidian Quash: From Mental Hygiene to Mental Health 1969–2011, Part 1 and Part 2.*

Dear Daughter:

I am ashamed of the United States of America. I am ashamed of California's criminal justice system. I am ashamed of my CONREP's past community outpatient therapy, and I am especially ashamed for my Napa State Hospital because they have *no shame*. However, I am *not* ashamed of myself, and I am not ashamed of you, Ms. Elaine Rose Hawaii, or any of our kind family. Kudos and praise for our exceptional achievements. You are an indispensable and very professional aide and daughter. When I just could *not* help myself financially here, thank you for being you and for helping me out so adequately.

Before part (2) is the same document photocopied, except for the rewritten *A Three-Part Discussion* and except for the 1960s date of the African American psychiatrist Dr. Vernon Indiana's suicide. I changed it to the more accurate 1970s date. Part 2 is all new. Please read all the four-paged Thursday, May 19, 2011, letter to the Honorable Judge Wyoming, in Part 2, about anger. *Dearest Elaine, especially read all of The Truth, the Whole Truth, and Nothing but the Helpful Truth on page 2 of the Thursday, May 19, 2011, letter, and let me know how you feel. Now I have nothing to hide. I can even say my disclosure is brilliant. And I can say your reading about you on page 2 will help me out of here and back to an outpatient without its previous "backstabbing peculating of the flesh."*

And furthermore, my good old friend Mr. Aruther McDonald has a good copy of *A Quotidian Quash: From Mental Hygiene to Mental Health 1969–2011, Part 1 and Part 2* to give to an attorney-at-law. He is to find me a winning attorney-at-law. This attached document of 132 pages is mostly my life's work against my 2011 mental health system. In conclusion, this letter comes to you saying (sometimes) a good confession, like this, is good for the soul. Good luck with your injured arm, and again, thank you!

Love,

Dad
Psychiatric patient

Mr. Dorian Gaylord Redus
Napa State Hospital
Ward T-15
2100 Napa-Vallejo Hwy.
Napa, CA 94558-6234
1(707)252-9988

Friday, October 7, 2011

The San Francisco Chronicle
901 Mission Street
San Francisco, CA 94103-2988
1(415) 777-1111

Please do find three letters attached: this new letter, the September 22, 2011, first letter addressed to the *San Francisco Chronicle* newspaper. And please also find the Wednesday, September 28, 2011, letter to my psychiatrist, Dr. Eric Florida.

Re: *A Quotidian Quash: From Mental Hygiene to Mental Health 1969–2011, Part 1 and Part 2.*

To Whom It May Concern:

I have a 132-page manuscript for you. I want the name of a nice person at the *San Francisco Chronicle* address above to send my written work to. Although, I first wrote to you two weeks ago, and I have not had any kind of follow-up contact by mail or by phone, I am still keeping my hopes of hearing from you alive. I have no internet service, and I am very technologically embarrassed in other ways. As a patient here, my only quash fighting option is computer word processing in a program suite new to me on a 3.5" floppy disk and printing my work on it in a lab half a mile away in another building. However, this is a path that all staff here at Napa State Hospital may accept, follow, and allow with their full approval and personal approbation. We have all their supports.

Should you need to, you should reach me by phone between the hours of 6:00 a.m. and 9:00 a.m., 12:30 p.m. and 2:30 p.m., after 4:45 p.m., or by mail at my address above. Thank you.

I trust that you will like these letters. I wrote the (two-and-one-half-paged) Wednesday, September 28, 2011, letter to my psychiatrist, Dr. Eric Florida, all by myself. *He refused to read it*, and nobody else has seen it except for you.

Please acknowledge, and please also forward a name to me of a very nice person at the *San Francisco Chronicle* newspaper that I may post my manuscript to as soon as possible. Thank you very much.

Respectfully submitted,

Mr. Dorian Gaylord Redus
Psychiatric patient

Mr. Dorian Gaylord Redus
Napa State Hospital
Ward T-15
2100 Napa-Vallejo Hwy.
Napa, CA 94558-6234
United States of America
1(707)252-9988

Monday, October 10, 2011

Dr. Steven W. Hawking
C/O Cambridge University
The Old School, Trinity Lane
Cambridge, CB2 1TN England

To Whom It May Concern:

    Thank you for receiving this letter. I am sixty-five years old. The eighteen-page document is one half of my life's work.
    In the 1990s, my mental health system called the part of this work that I had then my delusion about my universe. However, as my work was once summarily dismissed by some very nontherapeutic medical doctors now, I am keeping hope alive that you will free me by stopping the preponderating forensic experts here from considering my work so bizarre as to be unworthy of consideration.
    This is a last best hope—what American NFL (National Football League) television spectators call a game-winning Hail Mary touchdown pass. I can arrange a payment for any opinion given regarding my intellectual property, *A Three-Part Discussion*. Today, this is a tall order. As I write, it may be too tall. However, please do send me any written reaction to my written work.
    Please do also acknowledge your receipt of my work, *A Three-Part Discussion*.
    Thank you.

Respectfully submitted,

Mr. Dorian Gaylord Redus
Psychiatric patient

# A QUOTIDIAN QUASH: FROM MENTAL HYGIENE TO MENTAL HEALTH

Mr. Dorian Gaylord Redus
Napa State Hospital
Ward T-15
2100 Napa-Vallejo Hwy.
Napa, CA 94558-6234
1(707)252-9988

Tuesday, October 18, 2011

Joel R. Primack
Nancy Ellen Abrams
C/O Agent: Doug Abrams
106 Corinne Avenue
Santa Cruz, CA 95065
1(831)465-9565

To Whom It May Concern:

Thank you for receiving this letter. I am sixty-five years old. The eighteen-page document is one half of my life's work.

In the 1990s, my mental health system called the part of this work that I had then my delusion about my universe. However, as my work was once summarily dismissed by some very non-therapeutic medical doctors now, I am keeping hope alive that you will free me by stopping the preponderating forensic experts here from considering my work so bizarre as to be unworthy of consideration.

This is a last best hope—what American NFL (National Football League) television spectators call a game-winning Hail Mary touchdown pass. I can arrange a payment for any opinion given regarding my intellectual property, *A Three-Part Discussion*. Today, this is a tall order. As I write, it may be too tall. However, please do send me any written reaction to my written work.

When I peruse Joel R. Primack and Nancy Ellen Abrams's *The View from the Center of the Universe: Discovering Our Extraordinary Place in the Cosmos*, I like it very much.

Please do also acknowledge your receipt of my work, *A Three-Part Discussion*.
Thank you.

Respectfully submitted,

Mr. Dorian Gaylord Redus
Psychiatric patient

Mr. Dorian Gaylord Redus
Napa State Hospital
Ward T-15
2100 Napa-Vallejo Hwy.
Napa, CA 94558-6234
United States of America
1(707)252-9988

Friday, October 21, 2011

Dr. Lee Smolin
Perimeter Institute for Theoretical Physics
Waterloo, Ontario N2L 2Y5, Canada

To Whom It May Concern:

Thank you for receiving this letter. I am sixty-five years old. The eighteen-page document is one half of my life's work.

In the 1990s, my mental health system called the part of this work that I had then my delusion about my universe. However, as my work was once summarily dismissed by some very nontherapeutic medical doctors now, I am keeping hope alive that you will free me by stopping the preponderating forensic experts here from considering my work so bizarre as to be unworthy of consideration.

This is a last best hope—what American NFL (National Football League) television spectators call a game-winning Hail Mary touchdown pass. I can arrange a payment for any opinion given regarding my intellectual property, *A Three-Part Discussion*. Today, this is a tall order. As I write, it may be too tall. However, please do send me any written reaction to my written work.

Please do also acknowledge your receipt of my work, *A Three-Part Discussion*.

Thank you.

Respectfully submitted,

Mr. Dorian Gaylord Redus
Psychiatric patient

Mr. Dorian Gaylord Redus, Ward T-15
Napa State Hospital
2100 Napa-Vallejo Hwy.
Napa, CA 94558-6234
United States of America
1(707)252-9988

Monday, March 28, 2011

Dr. Lee Smolin
Perimeter Institute for Theoretical Physics
Waterloo, Ontario N2L 2Y5, Canada

Re: your October 28, 2010, *Nature* article "Space-time turn around" about Roger Penrose's book *Cycles of Time: An Extraordinary New View of the Universe*.

Dear Dr. Lee Smolin:

The deliciously absurd is usually easy to quash. However, your October 28, 2010, *Nature* article "Space-time turn around" about Roger Penrose's new book *Cycles of Time: An Extraordinary New View of the Universe* is most unusual. I obsecrate, beg, that you condescend and reply to me after reading pages 12 and 13 on my (STS) space-time sphere theory. Off the campus of my university, where I am a junior, my work has been summarily dismissed and called my "well-articulated delusion about the universe." I hope you are very happy to receive my undergraduate work, *A Three-Part Discussion*. Perhaps you have the time to read it and reply to me on my (STS) space-time sphere theory on its pages 12 and 13.

When I read over my work, *A Three-Part Discussion*, its RCTVU section seems to have a non sequitur or something queer about how author Dr. Brian Greene's "travel time effect" on pages 6 and 7 is connected to my observations on "space-time slowing" due to time dilation on pages 7 and 9. Please help me by shedding some much-needed light on their connection. Is it that time dilation causes Gracie's clock to run slow to George on page 6, and time dilation causes George's clock to run slow to Gracie on page 7 in the excerpt from Dr. Brian Greene's book *The Elegant Universe*? Here I am referring to the first, the second, and not the third paragraph of the excerpt.

I am keeping my hopes of hearing from you alive. I have no internet service here, and I am very technologically embarrassed in other ways. As a patient here, my only quash fighting option is computer word processing in a program suite new to me on a 3.5" floppy disk and printing my work on it in a lab half a mile away in another building. However, this is a path that all staff here may accept, follow, and allow with their full approval and personal approbation. We have their support.

Thank you.

Respectfully submitted,

Mr. Dorian Gaylord Redus
Patient

Mr. Dorian Gaylord Redus
Napa State Hospital
Ward T-15
2100 Napa-Vallejo Hwy.
Napa, CA 94558-6234 1(707)252-9988

                         Halloween, Monday, October 31, 2011

Dr. Alex Oklahoma, MD
Harvard University
Harvard Medical School Room: TMEC 244 260
Longwood Avenue Boston, MA
02115 1(617) 432-2159

Re: REDUS, DORIAN                   DOB: 05/19/46
    SC#: 88778                          PC 1026

*Subject:* a commonplace or quotidian quash anywhere is a quotidian quash everywhere.

*Legal Status:* I am a sixty-five-year-old African American male under a PC 187, murder. On August 9, 1974, I stabbed a woman I had known for six years to death. In October 1975, I was found not guilty by reason of insanity and remanded to Atascadero State Hospital. I was transferred to Napa State Hospital in 1982 (Atascadero and Napa state hospitals are in California), and I was released from Napa on to outpatient status in September 1988, under the supervision of San Francisco's CONREP (conditional release program). Finally, on October 1, 2009, I returned to Napa for keeping the angry voices I was hearing for months a secret. And on May 5, 2010, my outpatient status was revoked. Now I want to win a sanity hearing or trail this year, 2011.

To Whom It May Concern:

    Thank you for receiving these letters. I would like to send you my 132-page manuscript. I saw most of your 2002 TV show on "segregated medicine being an oxymoron." It was entitled "The Tuskegee Syphilis Study" or something like that, and it was sponsored by our Commonwealth Club of California and the California Healthcare Foundation. Actually, I saw most of it twice.

I am writing you now because your TV show helped me to express myself, and because my California hospital psychiatrist, Dr. Eric Florida, MD, refused to read my (attached) three-page letter to him. He said, "I don't have the time to read it." Without a nurturing reader, I fear that I will regress and trade all my dignity for a reinforced and hypocritical traducing too less than the long-term-care psychiatric patient and aborning author I was before my psychiatrist, Dr. Eric Florida, MD, refused to read my latest three-page letter to him.

My California psychiatrist, Dr. Eric Florida, MD, also refused to read my manuscript, which took me one whole year to write, while I was on his sex offender ward. Will you please ready my manuscript and my (attached) therapeutic three-page letter?

If you readily read between their lines about my therapy programs, then you will readily comprehend that the preponderating authorities here are legally oxymoronic according to my material manuscript.

In my manuscript, you will see prejudice and character assassination in issues like *Dorian is delusional, and he writes bizarre letters* and *I am delusional, and I write bizarre letters*. Where ad baculum is an inappropriate authority that in the past has intimately maintained that my two therapeutic stellar theories, STS (space-time sphere) and RCTVU (relativistic color television universe), are (de facto) my well-developed delusional theories on my autistic universe, as I am very sane, please read my very appropriate written work, and then draw your conclusion, supportive or otherwise, in a letter for my attorney-at-law.

When I can pay, as you read every page of my manuscript, please do not treat me as if the only thing I am good for is being an example of "good for nothing," because that will, very regrettably for me, make you in my court affairs "good for nothing."

I pray you will ask me for my manuscript, and if you should need to, you should reach me by phone between the hours of 6:00

a.m. and 9:00 a.m., 12:30 p.m. and 2:30 p.m., after 4:45 p.m. PST, or by mail at my address above.

Dr. Alex Oklahoma, MD, please do help me to increase my free will.

Thank you again.

Respectfully submitted,

Mr. Dorian Gaylord Redus
Psychiatric patient

Mr. Dorian Gaylord Redus
Napa State Hospital
Ward T-15
2100 Napa-Vallejo Hwy.
Napa, CA 94558-6234
1(707)252-9988

Monday, November 7, 2011

Department of Veterans Affairs
Oakland Regional Office
Oakland Federal Building
1301 Clay Street
North Tower Twelfth Floor
Oakland, CA 94612
1(800)827-1000 Ex 110

Re: REDUS, DORIAN GAYLORD

*Subject:* medicine, money and benefits, intransigent educational and racial prejudice, and also that a commonplace or quotidian quash anywhere is a quotidian squash everywhere.

*Legal Status:* I am a sixty-five-year-old African American male under a PC 187, murder. On August 9, 1974, I stabbed a woman I had known for six years to death. In October 1975, I was found not guilty by reason of insanity and remanded to Atascadero State Hospital. I was transferred to Napa State Hospital in 1982 (Atascadero and Napa state hospitals are in California). And I was released from Napa on to outpatient status in September 1988, under the supervision of San Francisco's CONREP (conditional release program). Finally, on October 1, 2009, I returned to Napa for keeping the angry voices I was hearing for months a secret. And on May 5, 2010, my outpatient status was revoked. Now I want to win a sanity hearing or trail this year, 2011.

To Whom It May Concern:

Thank you for receiving my (132-page) manuscript on psychiatric malpractice and these few letters. Thank you for your very helpful and timely payments of my VA benefits from approximately 1969 to 2011. Although, my pulse has since been a little high, thank you very, very much for my 2003 triple coronary artery bypass surgery and my dentures. Both done at Fort Miley in San Francisco by Chief of Cardiology Dr. Ratcliff, MD and (excellent student) Dr. Olmezova, DDS, respectively.

I am writing you now to officially thank the Department of Veterans Affairs for letting me write to you about receiving two to ten million dollars due to grievous malpractice by a Chief of Mental Hygiene at Fort Miley from 1970 to 1974, as it is still crucifying and crippling me in courtrooms and hospitals today. The abject psychiatrist is the late Dr. Donald Montana, MD.

When you pay me, please do not treat me as if the only thing I am good for is being an example of "good for nothing," because that will, very regrettably for me, make you in my court affairs "good for nothing."

I pray you will read my manuscript, and if you should need to, you should reach me by phone between the hours of 6:00 a.m. and 9:00 a.m., 12:30 p.m. and 2:30 p.m., after 4:45 p.m. PST, or by mail at my address above.

If you help me to increase my free will, then all I need is editing and publishing and payment.

Thank you again.

Respectfully submitted,

Mr. Dorian Gaylord Redus
Psychiatric patient

Mr. Dorian Gaylord Redus, T15
Napa State Hospital
2100 Napa-Vallejo Hwy.
Napa, CA 94558-6234
1(707)252-9988

Wednesday, November 9, 2011

MATIER & ROSS
The San Francisco Chronicle
901 Mission Street
San Francisco, CA 94103-2988
1(415)777-1111

To Whom It May Concern:

 There is (and has been) something terribly wrong with first, my mental hygiene, and second, my mental health system. Despite that, I feel any publicity will be therapeutic for me. Please do peruse all of my manuscript, and let me know how you feel about the publication of any of my well-documented issues. For example, when I think I should be given a "medication holiday" from all of my Risperidone 3 mg. psychotropic here in the safety of this secure hospital, quite inappropriately, the hospital seems to think my Risperidone 3 mg, psychotropic is as necessary as an absolutely appropriate (all-healing) panacea. In further example, I know that my mental health system caused my (apoplexy and my) 1990s divorce. The preponderating oxymoronic lies (in my court hearings and in my court trials) are an overwhelming juggernaut, over-mind, that lies (double entendre) above our laws.
 When I had reported some serious shortness of breath on Friday, June 24, 2011, and I reported very minor itchy feet late on Thursday night, June 30, 2011, the last day of June, I was seen for itchy feet. The very next day, the first day of July, Friday, July 1, 2011, and at sixty-five years of age, I was not seen for my serious shortness of breath until Wednesday, July 6, 2011. When I wrote my only liv-

ing brother, my unanswered letter gave me the serious shortness of breath. I could not breathe deeply because an attempt at a full breath made me feel like there was something very wrong with my heart.

Without a "medication holiday," all I can say is I am not a homosexual. However, I am here because I am queer, and I am queer because I am here. My mental health system is not (and has not been) therapeutic to me. My mental health system has (most ironically) promoted one harmful mentally filthy milieu for me after another—for decades.

Circa August 12, 1974, my August 9, 1974, murder was front-page news in that day's issue of *The San Francisco Chronicle*. That one day, the coverage was one column on the front page. Thank you.

Respectfully submitted,

Mr. Dorian Gaylord Redus
Psychiatric patient

## A QUOTIDIAN QUASH: FROM MENTAL HYGIENE TO MENTAL HEALTH

Mr. Dorian Gaylord Redus
Napa State Hospital
Ward T-15
2100 Napa-Vallejo Hwy.
Napa, CA 94558-6234
1(707)252-9988

Veteran's Day, Friday, November 11, 2011

Sally Tennessee
Executive Director
Napa State Hospital
2100 Napa-Vallejo Hwy.
Napa, CA 94558-6234

Re: REDUS, DORIAN            DOB: 05/19/46
    SC#: 88778            PC 1026

*Legal Status:* I am a sixty-five-year-old African American male under a PC 187, murder. On August 9, 1974, I stabbed a woman I had known for six years to death. In October 1975, I was found not guilty by reason of insanity and remanded to Atascadero State Hospital. I was transferred to Napa State Hospital in 1982, and I was released on outpatient status in September 1988 under the supervision of San Francisco's CONREP. Finally, on October 1, 2009, I returned to Napa for keeping the angry voices I was hearing for months a secret. And on May 5, 2010, my outpatient status was revoked. Now I want to win a sanity hearing or trail this year, 2011, and my public defender has not returned even one of my weekly telephone calls since January this year.

To Whom It May Concern:

    Thank you for receiving this correspondence. Thanksgiving is fast upon us, and I am expecting my two grown granddaughters and their mother, my daughter, at Napa's luncheon on Thursday, the seventeenth.

Thank you again for attending the recent veteran's celebration here. I was very mindful of you sitting in the audience, in a metal folding chair directly in front of me, at the celebration. I am also mindful that I need a much better mattress. In the audience, you look like your patients, like me, have hope that you will help us.

In the mid-1980s, I was a patient here on Ward Q3 and Q4. Way back then, as a younger male patient, I worked on Napa State Hospital's laundry truck delivering and picking up clean and soiled laundry. The wet soiled and dirty laundry bags that I had to pick up, all by myself, had to be held at my arm's length because they were wet with urine. Way back then, as a younger male patient, working on our laundry truck picking up our soiled laundry, I hurt and badly damaged my back.

*My hospital therapist's malpractice was that I was told in the mid-1980s that the x-rays, due to my great back pain, showed nothing. Much latter in 1988 (out of pocket), under the supervision of San Francisco's CONREP, I sought the help of a helpful chiropractic therapist. Her x-rays, due to my great back disability, showed a crushed disc caused my back pain, my disability, and showed that my crushed disc also caused my need for years of her services.*

Much more recently, here my back pain MRI of 9-16-11 in Sonoma's results were:

- No compression fracture
- Central disc protrusion—at L4-5

And my back pain x-ray of 8-10-11 at Napa State Hospital had the following results:

- Compression lumbar fracture at L1 and lower thoracic
- Osteoporosis
- Degenerative changes
- Cannot tell from when

Finally, I have one of the "new gray plastic foam mattresses" because of my back problems, which run in my family. *I need a more adequate and a more therapeutic mattress.*

While I am being so up close and personal, I have another (personal) problem that you look like you should help me with. Our (pun) psychiatrist, Dr. Eric Florida, MD, said, "I do not have time to read your letter." He said it to me on September 30, 2011, at my annual team conference. Without a nurturing reader, I fear that I will regress and trade all my dignity for a reinforced and hypocritical traducing too less than the long-term-care psychiatric patient and aborning author I was before our psychiatrist, Dr. Eric Florida, MD, refused to read my latest three-page 9-28-11 letter to him. Very reasonably, please read the attached copy of my three-page letter to Dr. Florida that I have sent to you for your help.

If you ever choose to also read my 132-page manuscript, *A Quotidian Quash: From Mental Hygiene to Mental Health Part 1 and Part 2*, then I have a nice spiral-bound copy that I am holding just for you. And moreover, if you ever choose to read it, please find: "When there is something all wrong and oxymoronic with my psychiatrist," and he "ain't hitting on nothing" but power relationships, then there is something psychotic looking about me. Sally, if you choose to read all of my manuscript, then you may also choose to attribute my many therapists' ways to malice or to attribute my many therapists' ways to incompetence knowing that I should (according to Napoleon Bonaparte on page 17 of *The Week* news media magazine for August 5, 2011) "Never ascribe to malice that which is adequately explained by incompetence."

Although I cannot get one for myself, I do feel that I need a much more therapeutic mattress.

Respectfully submitted,

Mr. Dorian Gaylord Redus
Psychiatric patient

Mr. Dorian Gaylord Redus,
Napa State Hospital
Ward T-15
2100 Napa-Vallejo Hwy.
Napa, CA 94558-6234
1(707)252-9988

Friday, November 18, 2011

Re: yesterday's discussion at Napa's Thanksgiving luncheon and your first call to my current Ward T-15 psychiatrist, Dr. Eric Florida, MD.

Please find enclosed one 11-11-11 letter to Napa's executive director and two letters to Dr. Eric Florida.

Dear Daughter:

This is not the twenty-four pages that we spoke of yesterday. They will come to you later.

You met my doctor yesterday at the luncheon. To telephone him, please, dial *1(707)253-5040*. That is Ward T-15's office and nursing-station. When they pick up, ask them to simply just connect you to Dr. Eric Florida in his nearby ward office. He may be very busy.

Thank you for receiving this important correspondence. Thank you for coming to yesterday's Thanksgiving luncheon. And (unfortunately), when you try to help me, you should prepare for and expect derision, being laughed at, as in the root word *deride*, or from my dictionary, to speak of or treat with contemptuous mirth.

I hope to see you and the girls on or around December 10, 2011, at the Family Support Group Holiday Luncheon here.

Love,

Dad
Psychiatric patient

Mr. Dorian Gaylord Redus
Napa State Hospital
Ward T-15
2100 Napa-Vallejo Hwy.
Napa, CA 94558-6234
1(707)252-9988

Saturday, November 19, 2011

Dr. Neil de Grasse Tyson
Princeton University
Princeton, New Jersey 08544
1(609)258-3000

Re: cosmology

*Subject:* My enclosed treatise, *A Three-Part Discussion.*

*Legal Status:* I am a sixty-five-year-old African American male under a PC 187, murder. On August 9, 1974, I stabbed a woman I had known for six years to death. In October 1975, I was found not guilty by reason of insanity and remanded to Atascadero State Hospital. I was transferred to Napa State Hospital in 1982 (Atascadero and Napa state hospitals are in California). And I was released from Napa on to outpatient status in September 1988 under the supervision of San Francisco's CONREP (conditional release program). Finally, on October 1, 2009, I returned to Napa for keeping the angry voices I was hearing for months a secret. And on May 5, 2010, my outpatient status was revoked. Now I want to win a sanity hearing or trail this year, 2011. And since I sent my California public defender attorney-at-law my cosmological treatise, *A Three-Part Discussion*, on January 11 this year, she has not returned even one of the weekly telephone calls that I placed to her.

To Whom It May Concern:

Thank you for receiving this letter. The eighteen-page document is one half of my life's work.

In the 1990s, my mental health system called the part of this work that I had then my delusion about my universe. However, as my work was once summarily dismissed by some very nontherapeutic medical doctors now, I am keeping hope alive that you will free me by stopping the preponderating forensic experts here from considering my work so bizarre as to be unworthy of consideration.

*This is a last best hope—what American NFL (National Football League) television spectators call a game-winning Hail Mary touchdown pass. I can arrange a payment for any opinion given regarding my intellectual property, A Three-Part Discussion. Today, this is a tall order. As I write, it may be too tall. However, please do send me any written reaction to my eighteen-page written work.*

I have more than one DVD from the Teaching Company's Great Courses series, and your *My Favorite Universe* is a favorite of mine. Currently, your 2002 *My Favorite Universe* is my most useful course. In 1972, I took an intrinsic summer seminar in cosmology at San Francisco's community college. City College of San Francisco's televised summer seminar of 1972, *Stellar Evolution: Man's Descent from the Stars*, was just classes and lectures by—disclaiming my spelling—Mr. Ray Bradbury, Dr. Geoffrey Burbidge, Dr. Edwin E. Salpeter, Dr. P. J. E. Peebles, Professor J. W. Schof, Dr. Sherwood Washburn, Dr. Melvin Calvin, Dr. Philip Morrison, Dr. Freeman Dyson, and Dr. Bernard M. Oliver of HP Company's "Project Cyclops."

Dr. Neil Tyson, this is my eighth inquiry. Four astrophysicists, three mental health doctors here at Napa State Hospital, and my inquiry to my school where I am a junior, San Francisco State University, and SFSU was the only one to answer my inquiry in writing.

Please acknowledge your receipt of my work, *A Three-Part Discussion.* And please send me any written reaction that you have to my written work. Perhaps you will elect to not acknowledge, and perhaps you will elect to acknowledge. And also, help me professionally and formally by engaging yourself in a friendly child's game of pin the tail on the donkey.

    Thank you.

Respectfully submitted,

Mr. Dorian Gaylord Redus
Psychiatric patient

Mr. Dorian Gaylord Redus
Napa State Hospital
Ward T-15
2100 Napa-Vallejo Hwy.
Napa, CA 94558-6234
1(707)252-9988

                                          Friday, December 2, 2011

Cheryl H. Arkansas
Attorney-at-Law
214 Duboce Avenue
San Francisco, CA 94103
1(415)431-0425

re: REDUS, DORIAN

*The Subject:* urgent.

*Legal Status:* I am a sixty-five-year-old African American male under a PC 187, murder. On August 9, 1974, I stabbed a woman I had known for six years to death. In October 1975, I was found not guilty by reason of insanity and remanded to Atascadero State Hospital. I was transferred to Napa State Hospital in 1982, and I was released onto outpatient status in September 1988, under the supervision of San Francisco's CONREP. Finally, on October 1, 2009, I returned to Napa for keeping the angry voices I was hearing for months a secret. And on May 5, 2010, my outpatient status was revoked. Now I want to win a sanity hearing or trail this year, 2011. And since I sent you, my public defender attorney-at-law, my cosmological treatise, *A Three-Part Discussion*, on January 11 this year, you did not return even one of the weekly telephone calls that I placed to you until yesterday.

To Whom It May Concern:

Thank you for receiving this letter. I am still fine, and I am still living on here at this hospital for the criminally and mentally insane. In the media, TV, or the *San Francisco Chronicle*, assert in these words or your own choice of words:

> There is a continuing homosexual rape afoot at Napa State Hospital involving hundreds of consensual oral copulations induced by mandated ingestions and injections of abject psychiatric drugs, as if they are medications. So the main victim, Mr. Dorian Gaylord Redus, in these consensual rapes due to abject psychiatric drugs (thirty-sexy years ago that still threaten him at Napa today) needs cosmologists to help discover and adduce his personal astrophysics theory: RCTVU (relativistic color television universe). So the California mental healthcare system won't make him an infamous intellectual inpatient baby raping child molester—just because his psychiatrists cannot do their psychedelic jobs and work. Intrinsically Gaylord's many psychiatrists have been his frightful clowns for forty years.

Respectfully submitted,

Mr. Dorian Gaylord Redus

Mr. Dorian Gaylord Redus
Napa State Hospital
Ward T-15
2100 Napa-Vallejo Hwy.
Napa, CA 94558-6234
1(707)252-9988

Friday, December 9, 2011

Deidre A. Defreese
Senior Student Support Services Specialist
Disability Programs and Resource Center
San Francisco State University
1600 Holloway Avenue
Student Services Bldg. 110
San Francisco, CA 94132
Appt: (415)338-2472
    fax: (415)338-1041
    (415)338-6356 direct line

Re: my personal psychological research done here while a junior at San Francisco State University.

*Subject:* my enclosed treatise, *A Quotidian Quash: From Mental Hygiene to Mental Health 1969–2011 Part 1 and Part 2.*

*Legal Status:* I am a sixty-five-year-old African American male under a PC 187, murder. On August 9, 1974, I stabbed a woman I had known for six years to death. In October 1975, I was found not guilty by reason of insanity and remanded to Atascadero State Hospital. I was transferred to Napa State Hospital in 1982 (Atascadero and Napa state hospitals are in California). And I was released from Napa onto outpatient status in September 1988, under the supervision of San Francisco's CONREP. Finally, on October 1, 2009, I returned to Napa for keeping the angry voices I was hearing for months a secret. And on May 5, 2010, my outpatient status was revoked. Now I want to win a sanity hearing or trail this year, 2011. And since I sent my California public defender attorney-at-law my treatise, *A Three-Part Discussion*, on January 11 this year, she did not return even one of the weekly telephone calls that I placed to her until Thursday, December 1, 2011.

To Whom It May Concern:

Thank you for receiving all of this stuff: my forth and most important letter to you, these thirteen one and two-page therapeutic letters to others, my latest eighteen-page treatise on my cosmology, and my main manuscript, *A Quotidian Quash*.

I need poetic justice, admonishment for my court officers malice and intransigent incompetence. And I also need poetic justice, a reward for my virtue. If you cure me by securing the careful reading of this long letter, these few pages of documents, my latest cosmology essay, and finally, if you could appropriate faculty at SFSU to look at my whole manuscript, then we may make my California mental health system envious.

Every day here at Napa State Hospital, their envy causes regressive "repressed disagreements" in hospital staff and patients alike. Far too often, preponderating court officers only see the pathological in me, their nice patient, and they take their patients qualities and possessions. Moreover, my mental health system's jealousy, paranoid fears regarding the bitter rivalry to queer me into submission to them or queer me to death as a vain martyr, in courtrooms, is all too obvious to my readers of my manuscript.

One may solve my problem knowing that I am a teddy bear with a tenable TV theory that threatens my mental health system (only) if they oppose it and call it a pathological psychological disease.

I confess when my preparation kisses my opportunity and when there is poetic justice, my psychiatrists and also my attorneys-at-law may be admonished here or in San Francisco where my nuclear family—mother Vivian, father Caleb, me, and my three siblings—got our first TV when I was a small boy. Especially when there was poetic justice in TV westerns with some foiled machination or evil scheme.

As children, we would exclaim in psychological excitement, "The good guys are coming!" Now San Francisco State University is the rectitude. So I sorely need SFSU to help me up to a more workable level. In 2009, I left SFSU as a junior with no feedback on my manuscript's issues.

There are so many wrongs evidenced in the many effective problem-solving letters to my current and preponderating authorities. I don't even know what to ask you to do for me to help me up. However, I have some suggestions: in part, I fit in here, but I do not belong here as just a poorly informed patient. I am daft here that your short July 21, 2011, memo of reply to my Monday, March 28, 2011, letter to you is the only written reply to any of the effective letters in my 132-page treatise, *A Quotidian Quash: From Mental Hygiene to Mental Health 1969–2011 Part 1 and Part 2*.

For some reason unknown to me, two fellow patients—first, one male, and second, one female—jumped out of their San Francisco dwelling's window, and they both nearly died. I had had a long-standing platonic relationship with the both of them. The three of us were in the same outpatient treatment program, and so we were affiliated/close.

*The only San Francisco State University class that I finished was Dr. Paap's psycholinguistics class in, as I recall, 2009. I finished Dr. Paap's psycholinguistics class with a B+ (instead of the A on my final) because of the horrible personal tragedy suffered by a close female friend just days before one of Dr. Paap's midterms. I believe I got an insurmountable D on that one midterm exam, no doubt due to my two friends' window tragedies.*

All my psychiatrists and all my attorneys-at-law have been just atrocious. Their nontherapeutic oxymoronic affiliations and de facto homosexual rape due to two of their atrocious abject psychiatric drugs (twenty-five years ago), their one mentally filthy milieu after another, and their oxymoronic abandonments have all been covered up until I think a quotidian quash anywhere is a quotidian quash everywhere.

I need to give and to receive therapy that is not the oxymoronic, crucifying, and crippling brand due to appropriating power relationships described in my enclosed treatise, *A Quotidian Quash:*

## A QUOTIDIAN QUASH: FROM MENTAL HYGIENE TO MENTAL HEALTH

*From Mental Hygiene to Mental Health 1969–2011 Part 1 and Part 2.* Please do help me by having my treatise: *A Quotidian Quash*, carefully read by some appropriate people like you who truly may help me. There are so many rubberneckers here and only alligators watching my back. Although, I first wrote to you some months ago, and I have only had your few lines of follow-up contact by mail or by phone, I am still keeping my hopes of hearing from SFSU alive. I have no internet service here, and I am very technologically embarrassed in other ways. As a patient here, my only quash fighting option is my computer word processing in a program suite new to me on a 3.5" floppy disk and printing my work on it in a lab half a mile away in another building. However, this is a path that all staff here may accept, follow, and allow with their full approval and personal approbation. We have their support.

> The world is a dangerous place. Not because of the people who are evil, but because of the people who don't do anything about it. (Albert Einstein in *The Week* for 11-25-11, page 23)

My own evaluative response regarding this, my new enclosed work, is "under a cloud" of quash, and all I think and feel is God is within us all, sometimes within one as one in their daily life, and I do not want the devil, even sometimes, in me as me in my daily life, so I obsecrate, beg, for your guidance.

My own conclusion is that here the putative is just to be punitive, and dignity draining and destroying regressive "repressed difficulties and disagreements" afflict all, severely worsening and aggravating Napa State Hospital's patients and staff alike…

Again, thank you for your help, your time, and your considerations. Have a nice day.

Respectfully submitted,

Mr. Dorian Gaylord Redus
Psychiatric patient

Mr. Dorian Gaylord Redus
Napa State Hospital
Ward T-15
2100 Napa-Vallejo Hwy.
Napa, CA 94558-6234
1(707)252-9988

Friday, December 9, 2011

Virginia Illinois
Social Worker
Napa State Hospital
Ward T-15
2100 Napa-Vallejo Hwy.
Napa, CA 94558-6234

Re: what you asked for at our chat yesterday.

To Whom It May Concern:

    11-02-11—I learned of a letter to my court written in May this year, and I requested the May letter Dr. Mexico told me about, 11-02-11. And also in the same request, I asked for a 10-07-09 report I needed by Dr. Dakota done on my admissions ward.
    11-16-11—I signed to pay $2.40 from my trust account for twenty-four pages to be copied and delivered to me here on Ward T-15 as soon as possible.
    11-21-11—I gave you the form that the unit supervisor and I signed, and you said that you sent it off.
    I personally purport, posit, and I behest that if I need a "medication holiday" and then a release without CONREP, the current plan to give me nontherapeutic Consta injections at an augmentation to 6 mg. and then CONREP receives an improvement needed, from me, as far as my current plans go. It is dishonest "people pleasing" on my part to agree on nontherapeutic Consta injections in compromise

just to get the possibility of CONREP, and the choice is quite simply just against my better judgment.

Here, Dr. Eric Florida's judgment for Consta then CONREP is demonic to me. I need, first, a "medication holiday" here, and then second, more of my free will (accordingly) and less of my psychiatrist's free will. Where is my no? Where are my rights?

Thank you.

Respectfully submitted,

Mr. Dorian Gaylord Redus
Psychiatric patient

PS I saw Cheryl H. Arkansas, my attorney-at-law, today at Napa's visiting center, and we want Dr. Florida to start the Consta injections.

Mr. Dorian Gaylord Redus
Napa State Hospital
Ward T-15
2100 Napa-Vallejo Hwy.
Napa, CA 94558-6234
1(707)252-9988

Friday, December 16, 2011

Christopher A. Idaho, LCSW
Community Program Director
Anka Behavioral Health Services
Golden Gate Conditional Release Program
350 Brannan Street, Suite 200
San Francisco, CA 94107
1(415)222-6930

Re: REDUS, DORIAN GAYLORD and why I kept my angry voices a secret for eight months, from February to September 2009.

*Legal Status:* I am a sixty-five-year-old African American male under a PC 187, murder. On August 9, 1974, I stabbed a woman I had known for six years to death. In October 1975, I was found not guilty by reason of insanity and remanded to Atascadero State Hospital. I was transferred to Napa State Hospital in 1982, and I was released onto outpatient status in September 1988, under the supervision of San Francisco's CONREP. Finally, on October 1, 2009, I returned to Napa for keeping the angry voices I was hearing for months a secret. And on May 5, 2010, my outpatient status was revoked. Now I want to win a sanity hearing or trial this year, 2011. And since I sent my public defender attorney-at-law my cosmological treatise, *A Three-Part Discussion,* on January 11 this year, she did not return even one of the weekly telephone calls that I placed to her until Thursday, December 1, 2011. However, Friday, December 9, Cheryl H. Arkansas visited me here at Napa's visiting center.

To Whom It May Concern:

Thank you for receiving this letter. All my nice things from my old San Francisco sunset district apartment are all safe locally in storage until May 2012. My loving (only) daughter, Ms. Elaine Rose Hawaii, has recently seen to all my things with my friend's help finding the right place to store all of it.

I kept secrets from CONREP because CONREP kept secrets from me. What was the real reason I was in sex offender treatment for many years? And why was it for so many years? When everything is covered in my (enclosed) Friday, January 7, 2011, letter to Napa's Dr. Mexico, PhD, and when I had shared all of it verbally before my sex offender treatments even began in 2005, how are my sex offender treatments due to any of my past, present, or possible future sexual situations, actions, or behaviors?

I kept secrets from you, CONREP, because you were the child molesters without a cause. Please just see my (enclosed) letter to my current psychiatrist here, Dr. Eric Florida, dated Monday, June 13, 2011. And let it show you your bad, not my bad.

The secrets were linked to my anger. Mr. Christopher A. Idaho, you, my CONREP, had daft and dicey nontherapeutic goals and directions for me that I chose to be angry about and secretive about. Whenever I find anger in the (enclosed) Monday, November 15, 2010, letter, especially as epitomized in my "Four issues that have made me angry in the recent past," which CONREP most probably does not yet even want to talk about, those "Four issues that have made me angry in the recent past" exasperate me, have frustrated me, and they still vex me enough (to want to but not need) to respond to our current unequal "power relationship" by choosing to keep secrets from you, my CONREP and my Department of State hospitals.

The "four issues" are on page 6 of my (enclosed) November 15, 2010, six-page letter to my current public defender, Cheryl H. Arkansas. Nevertheless, my sharing and self-disclosure was so pejorative that it was bad for me, and it was good for nobody.

So I kept my secrets because you had too much information from me due to my self-disclosure. And furthermore, you could lie on me in court. You have discretion, and you are going to make up your own minds based on how much of my own free will you decide that I deserve, so you tell me why I kept my sanity secret.

---

It is more true than false that CONREP may one, domineer me *in saecula saeculorum* for ever and ever; two, allow a more respectful mutual dominion; or three, start arm-twisting to an "intellectually dishonest" Armageddon.

The following is what first, you, and second, I, have previously said about some of my anger.

3) Anger

> Mr. Redus is unable to moderate his anger and has exhibited signs of uncontrollable anger and rage during his instant offense. At the time, he believed that his girlfriend/victim was being unfaithful and that she wanted to harm him. In the past, he has gotten into physical altercations with his girlfriend, which has led to him striking her. Mr. Redus was initially in denial about being angry with his daughter for lending his money to someone and not being able to get repaid. He did not admit that he was angry until he was brought before the treatment team (on September 28, 2009) to discuss about his anger which manifested in having intrusive thoughts about hurting others. Although Mr. Redus may recognize some of his warning signs when he gets angry: irritability and agitation, thoughts about violence, his face turns red, he starts becoming delusional, and starts to have homicidal ideation, nevertheless, he was not forthcoming about his intrusive thoughts. He kept his anger to himself

and minimized his warning signs for 8 months before finally telling staff. This is an area where he needs to address his issues and focus on being honest with staff. (From page 5 of a six-page letter from you, CONREP, to the Honorable Judge Wyoming dated Monday, April 12, 2010)

## The Truth, the Whole Truth, and Nothing but the Helpful Truth

3) Anger

CONREP said, "Mr. Redus is unable to moderate his anger and has exhibited signs of uncontrollable anger and rage during his instant offense." That is a ramifying lie in print by liars who lie (double entendre) above the law. Really, I posit, due to my superior range of intelligence, my personal anger management is my forte or strong point. My CONREP said that, "At the time, he believed that his girlfriend/victim was being unfaithful and that she wanted to harm him." After my 1969–1974 "girlfriend/victim," Ms. Edna Ella Robenson, threw me out of our little in-law love cottage across the street from City College of San Francisco, I visited her, and at times, I saw her with another young man. She told me that she wanted to have sex with her visiting young man, student Mr. Lee Lenard. So for me, Edna was freer than she was unfaithful to me. Moreover, within the first few months (of meeting her), Edna became suicidal once and also attacked me with our kitchen knife, which happened many more times during the six years we were in our doctor-advised relationship. Sometimes, she attacked me with our kitchen knife, and she also called San Francisco Police to our small apartments and our little backyard in-law love cottage. Furthermore, when I was visiting her just before I took her life on the worst morning of my life, I found her on the telephone, in her sheer off-white bra and panties. And after Edna got off the telephone, she said, "That was the police. They are getting me a gun." Then Edna said she was going to kill me with the gun—expletive deleted. All of that puts a different light on CONREP's "At the time, he believed that his girlfriend/victim was being unfaithful and that she wanted to harm him." CONREP's casuistry continues with CONREP's "In the past, he has gotten into physical altercations with his girlfriend, which has led to him striking her." Because I thought and felt she was "flirting with death" and needed a fair game, one time, sort of, "warning shot across her bow," I struck her one time—once. What does really anger me and get my goat is CONREP's statement:

> Mr. Redus was initially in denial about being angry with his daughter for lending his money to someone and not being able to get repaid. He did not admit that he was angry until he was brought before the treatment team (on September 28, 2009) to discuss about his anger which manifested in having intrusive thoughts about hurting others.

I did not feel anger at my daughter, Ms. Elaine Rose Hawaii, because I more thought and felt I should choose to be protective than I felt I should choose to feel angry with her. Elaine was putting my youngest granddaughter through the University of San Francisco, a very auspicious undertaking at an expensive university for a single mother of two. My youngest granddaughter graduated in May 2011, and that is, obviously to me, more important to me than the $24,000 my daughter stole from my personal savings account. Besides all of that, to date, my daughter has paid me back, to date, probably $10,000. Besides all of that, I am angry at CONREP's casuistry because their casuistry used my supportive and loving extended family member as a "scapegoat."

> Although Mr. Redus may recognize some of his warning signs when he gets angry: irritability and agitation, thoughts about violence, his face turns red, he starts becoming delusional, and starts to have homicidal ideation, nevertheless, he was not forthcoming about his intrusive thoughts. He kept his anger to himself and minimized his

warning signs for 8 months before finally telling staff. This is an area where he needs to address his issues and focus on being honest with staff.

This is subtle, specious, and harmful "scapegoat" reasoning that is, in its own essence, misleading: rationalizations. CONREP is unjustly devising their own self-satisfying false reasons for my supposed behavioral anger with my daughter, Ms. Elaine Rose Redus. CONREP's "this is an area where he needs to address his issues and focus on being honest with staff" is a statement that makes me recall my "Four issues that have made me angry in the recent past." This is all in my *A Quotidian Quash: From Mental Hygiene to Mental Health 1969–2011 Part 1 and Part 2*.

---

I personally purport, posit, and I behest that if I need a "medication holiday" and then a release without CONREP, the current plan to give me nontherapeutic Consta injections at an augmentation to 6 mg., and then CONREP receives an improvement needed from me, as far as my current plans go. It is dishonest "people pleasing" on my part to agree on nontherapeutic Consta injections in compromise just to get the possibility of CONREP, and the choice is quite simply just against my better judgment.

Here, Dr. Eric Florida's judgment for Consta then CONREP is demonic to me. I need, first, a "medication holiday" here. And then second, more of my free will (accordingly) and less of my psychiatrist's free will. Where is my no? Where are my rights?

Thank you.

Respectfully submitted,

Mr. Dorian Gaylord Redus
Psychiatric patient

PS—I saw Cheryl H. Arkansas, my attorney-at-law, 12-09-11, at Napa's visiting center, and we want Dr. Florida to start the Consta injections.

Mr. Dorian Gaylord Redus
Napa State Hospital
Ward T-15
2100 Napa-Vallejo Hwy.
Napa, CA 94558-6234
1(707)252-9988

Saturday, December 17, 2011

National Alliance for the Mentally Ill
3803 Fairfax Drive
Arlington, VA 22201
1(703)524-7600
NAMI Help Line 1(800)950-6264

Re: REDUS, DORIAN GAYLORD

*Legal Status:* I am a sixty-five-year-old African American male under a PC 187, murder. On August 9, 1974, I stabbed a woman I had known for six years to death. In October 1975, I was found not guilty by reason of insanity and remanded to Atascadero State Hospital. I was transferred to Napa State Hospital in 1982, and I was released onto outpatient status in September 1988, under the supervision of San Francisco's CONREP. Atascadero and Napa state hospitals are in California. Finally, on October 1, 2009, I returned to Napa for keeping the angry voices I was hearing for months a secret. And on May 5, 2010, my outpatient status was revoked. Now I want to win a sanity hearing or trial this year, 2011. And since I sent my public defender attorney-at-law my cosmological treatise, *A Three-Part Discussion*, on January 11 this year, she did not return even one of the weekly telephone calls that I placed to her until Thursday, December 1, 2011. However, Friday, December 9, Cheryl H. Arkansas visited me here at Napa's visiting center.

# A QUOTIDIAN QUASH: FROM MENTAL HYGIENE TO MENTAL HEALTH

To Whom It May Concern:

Thank you for receiving this letter.

When I corrected about one-third of the written and preponderating misstatements in some court documents from California's Department of State hospitals to some of my San Francisco judges, they still continue to include the misleading misstatements. Videlicet, they called my late VA psychiatrist Dr. Donald Montana, MD, who was the Veterans 1970s Chief of Mental Hygiene in San Francisco, even though he malpracticed as a Veterans Administration doctor, my private psychiatrist.

When I had reported that he had told me (once or twice) in his VA office to "get a gun," they wrote at a preponderating "intellectually dishonest" subterfuge or devise on a Department of State hospital's legal court document: "He [Dorian Redus] had informed his psychiatrist at the time that his girlfriend was having an affair. He heard his psychiatrist respond, 'Get a gun and shoot her.'" My victim in my 1974 homicide was Ms. Edna Ella Robenson.

I need two doctors, who my California judge will listen to, to do three legal things. One, read about 150 pages of my writing. Two, see me here at Napa State Hospital's official visiting center to discuss my case with me and evaluate me. And three, I need them to legally and officially advise my court in San Francisco (I hope of my sanity) by also working with my attorney-at-law.

The preponderating oxymoronic lies in my court hearings and in my court trials due to my homicide thirty-eight years ago and due to my hundreds of consensual homosexual rapes approximately twenty-five years ago, they, are an overwhelming juggernaut, overmind, that lies (double entendre) above our laws. Videlicet, they recently wrote, "There is no known history of emotional or sexual abuse" on page 8 of a May 16, 2011, letter to the Honorable Master Calendar Judge San Francisco's County Superior Court by Dr. Eric Florida, MD, my current psychiatrist here at Napa State Hospital.

I have family here who have forgiven me. Through my essential bound photocopied manuscript of 132 pages, *A Quotidian Quash: From Mental Hygiene to Mental Health 1969–2011 Part 1 and Part*

2, I have forgiven myself and garnered a desire to get a life outside of California's new Department of State hospitals.
 Thank you.

Respectfully submitted,

Mr. Dorian Gaylord Redus
Psychiatric patient

Mr. Dorian Gaylord Redus
Napa State Hospital
Ward T-15
2100 Napa-Vallejo Hwy.
Napa, CA 94558-6234
1(707)252-9988

Sunday, Christmas Day, December 25, 2011

Dr. Eric Florida, MD
Ward T-15
Psychiatrist
Napa State Hospital
2100 Napa-Vallejo Hwy.
Napa, CA 94558-6234

Re: a "medication holiday" after an immediate lowering of my psychotropic Risperidone 3 mg.

To Whom It May Concern:

 Thank you for receiving this letter. Because it will increase my trust in them, I prefer a "medication holiday" from California's Department of State hospitals at this time in my Ward T-15 therapy. All the quotes in this letter are from page 11 of a report signed on October 12, 2009, and are from testing done on October 7, 2009, by Dr. Christina Dakota, PhD.
 As I am obviously a victim of many consensual homosexual rapes due to abject psychiatric drugs (twenty-five years ago) and therefore, I have a fear and a paranoia of punishment when I even ask, may I please be freed of all the abject with an immediate lowering of my psychotropic Risperidone 3 mg. medication until I may have a "medication holiday" maintained according to my improvement?
 I am not a loner, but please note:

> When able to read to himself, Mr. Redus
> did quite well deciphering the meaning of sen-

> tences (Sentence Comprehension Standard Score: 125, 95% chance actual score is between 116–132, Above Average-Upper Extreme range, >12.9 grade equivalent).

That the highest score here is when working alone comes from homosexual rape and its insecurities. The rape comes from my horrible drug reactions. The drugs came from my Atascadero psychiatrist and my Napa psychiatrist. The psychiatry came from my attorney-at-law, Hiram E. Smith, of the 1970s, San Francisco law offices of White Badman & Smith. Hiram E. Smith came from the redoubtable repartee, Mr. Willie L. White Jr., and I next came to feeling like killing people in the early 1980s because of my two horrible Department of State hospital's abject psychiatric medications. But I chose consensual homosexuality instead in the two all-male patient psychiatric environments. So any current psychotropic regimen, plan, or goal other than "medication holiday" is a bad regimen, plan, or goal.
I just need to be free to know what I do is right.

> The Wechsler Abbreviated Scale of Intelligence (WASI) utilizes age-based norms of T scores converted to deviation IQs with a mean of 100 and a standard deviation of 15. Mr. Redus obtained an estimated Verbal IQ of 120, 91st percentile, 95% chance actual estimated Verbal IQ is between 114-125, High Average-Superior range. He obtained an estimated Performance IQ of 121, 92nd percentile, 95% chance actual estimated Performance IQ is between 115-126, High Average-Superior range. Mr. Redus obtained an estimated Full Scale IQ of 124, 95th percentile, 95% chance actual estimated Full Scale IQ is between 119-128, Superior range.

So it seems that I cannot just trust California's Department of State hospitals unless they first trust me and they justly decide to give me a nice "medication holiday" from my Risperidone 3 mg. medication at this time in my therapy.

Because they all refused me my most appropriate psychotropic medication, I should not have trusted the Veterans Administration forty years ago, and because they all trust the Veterans, I cannot trust any (none) of my mental health psychiatrists since the VA.

In the safety of our secure treatment area, please lower my psychotropic Risperidone 3 mg. medication until I may have a "medication holiday" maintained according to my improvement as soon as possible.

Thank you.

Respectfully submitted,

Mr. Dorian Gaylord Redus
Psychiatric patient

Mr. Dorian Gaylord Redus
Napa State Hospital
Ward T-15
2100 Napa-Vallejo Hwy.
Napa, CA 94558-6234
1(707)252-9988

Monday, January 30, 2012

Cheryl H. Arkansas
Attorney-at-Law
214 Duboce Avenue
San Francisco, CA 94103
1(415)431-0425

*Regarding my treatise, A Three-Part Discussion, among other things.*

*Legal Status:* I am a sixty-five-year-old light-skinned African American male under a PC 187, murder. On August 9, 1974, I stabbed a woman I had known for six years to death. In October 1975, I was found not guilty by reason of insanity and remanded to Atascadero State Hospital. I was transferred to Napa State Hospital in 1982, and I was released onto outpatient status in September 1988, under the supervision of San Francisco's CONREP. Finally, on October 1, 2009, I returned to Napa for keeping the angry voices I had been hearing for months a secret. And on May 5, 2010, my outpatient status was revoked. Now I want to win a sanity hearing or trial this year, 2012. And after I sent you, my public defender attorney-at-law, my cosmological treatise, *A Three-Part Discussion*, on January 11, 2011, you did not return even one of the weekly telephone calls that I placed to you until Thursday, December 1, 2011. However, Friday, December 9 last year, you visited me here in Napa's visiting center.

To Whom It May Concern:

Thank you for receiving this our first correspondence this new year. The Department of State hospitals is sometimes extremely mean to me. They have sometimes been party to and participated in my "cruel and unusual punishment." And moreover, they often prevaricate on my august issues in high places (courts) where they preponderate. Either as I say, I care about myself and others, or as our Department of State hospitals have far too often said, I am a danger to myself and others. So I need you as my attorney to stop them for assuming and avoiding or "begging the questions" before my court in my many forensic affairs, especially at my next PC 1606 or PC 1608 hearing or trial.

Whereas their consequences are not impossible or absurd, the past interdisciplinary and dishonesty argument over my two cosmology theories was an unjust reductio ad absurdum or a weak disproof of my RCTVU (relativistic color television universe) and its STS (space-time sphere) cosmology theories by reducing them to absurdities. When they are carried to their logical conclusion, as I have done in my (enclosed) treatise, they are, to me, a good idea and a valuable intellectual property.

I do not feel or think that mimickers should ridicule my current work and give my current work the disabling reductio ad absurdum diagnosis that they gave it in the past as they laughed at me with derision. I feel and also think my work should be emulated by my equals who, with their greater free will and environmental resources, may rise up against my past self and my past "court officers," having risen up against my adversarial side with my latest treatise, *A Three-Part Discussion Including Part Four*. They may (next) imitate me and excel because of my work as they advise my next court. There are one or two fellow patients here on Napa's Ward T-15 who seem to understand some of my work, but I did all the seminal work myself. What are my options?

One of the two things that caused my main major decompensation (for forty years) is epitomized by the following segue:

> "The stone which the builders rejected as worthless turned out to be the most important of all" (Matt. 21:42).

Unless the second (temporal) wrong is to decide or determine what or who was wrong in the first place, two wrongs do not make a right. If my psychiatrists diagnose that my treatise, cornerstone, is a pathetic pathology or malfunction, then how will it change my treatment and my family? To reveal the stone, at this time, I proffer that we need all the formal, the official, and the authoritative opinions, opine, and wicked wit of all the cosmologists and astrophysicists we may be fortunate enough to afford and lure into court for my next PC 1606 or 1608 hearing or trial in San Francisco, California.

To rule out a problem from information relationships in the secure treatment area of this state hospital, a "medication holiday" was indicated. But I have been medicated here on Napa's Ward T-15, its best open ward, since September 15, 2010, and my recently requested "medication holiday" from my low dose of psychotropic medication for schizophrenia was denied on January 18, 2012, in utter disregard of my excellent behavior and recent election to Ward T-15's ward government as its vice president. Whereupon, I said, "If it is inevitable, then I want the medication change to Consta injections as soon as possible." And then my psychiatrist, social worker, and psychologist all laughed as I laughed.

Regret for the things we have done is tempered by time. It is the regret for the things we did not try that is inconsolable and making my side forlorn without an adequate attorney-at-law. In *The Week* (news media magazine) of 10-21-11, on page 35 in an obituary regarding Steve Jobs, I read, "He once said that taking LSD during the 1970s was one of the two or three most important things he had done in his life." What is wrong here? We have a choice. One of my outstanding roommates in my four-man dormitory has had a putative IQ of 72. However, he regularly bench-presses over 200 lb. At the time of their conception, the father of my two grown granddaughters bench-pressed 555 lb. If one heard my granddaughter's father (for some reason) bench-pressed 300 lb., then one might think that his "medication" had benefited him and given him strength, just like one might think my high average superior range IQ goes with my initial 1-27-12 Consta (medication) injection. I really do understand my prescribing psychiatrist's situation, feelings, and motives.

Dr. Eric Florida is covering up for two other prescribing physicians who prescribed me two other abject medications that, on the all-male hospital ward environments of the 1980s, were an agent of homosexual rape. It is all casuistry—misleading and self-serving rationalization. Today, I have empathy for all my many roommates.

As to my Prolixin and its homosexual raping due to Dr. Wiggly's abject psychiatry at Atascadero State Hospital c. 1981, and as to my Haldol and its homosexual raping due to Dr. Marshal Arizona's abject psychiatry here at Napa State Hospital c. 1983, when their consequence, rape, is not impossible or justly absurd, it is more reductio ad absurdum to preponderate here, in court, and at CONREP by reducing my allegations of rape to absurdities. According to some psychiatrists, LSD and psychiatry are really and unjustly tantamount to rape. Whatever, my allegations of homosexual rape are a good idea. They are not absurdities as much as they are unjustly nullified by a reductio ad absurdum argument.

According to my *Collier's* 1961 Funk & Wagnalls Company standard dictionary:

> *reductio ad absurdum* (ri-duk' she o)
> Literally, reduction to an absurdity; disposal of a proposition by showing that its logical conclusion is absurd; also, proof of a proposition by showing its contradictory to be absurd.

According to my memories from my grammar school, Abraham Lincoln said something like "You can fool some of the people all of the time and all of the people some of the time, but you can't fool all of the people all of the time." When it comes to fooling and to guiding people with the evil and the good of reductio ad absurdum argument, there are two key elements or ways to make the reduction to absurdity work—fooling with the logical conclusion and guiding people to the contradictory absurdities.

To fool some of the people all the time, the Department of State hospitals has killed all of its public psychedelic experiments in its hospitals by sharing their data on bad trips on LSD. This is disposal of a proposition by showing that its logical conclusion is absurd. There is also, as I have done above, the de facto proof of a proposition by showing its contradictory to be absurd. Videlicet, I have included guiding data on the billionaire Steve Jobs and included (here) that *The Psychedelic Experience* by Timothy Leary et al. does advise that the psychedelic experience is like a "cosmic TV." This is guiding proof of a proposition by showing its contradictory to be absurd, because, according to Abraham Lincoln, "You can't fool all of the people all of the time." Contradicting a billionaire is very absurd, and I like my cosmic TV theories, which I had before I perused *The Psychedelic Experience* by Timothy Leary et al.

To fool some of the people all the time in San Francisco County Superior Court, my past CONREP killed my cosmology by advising that my RCTVU (relativistic color television universe) and my STS (space-time sphere) theories are just my autistic delusions about my universe. This was disposal of my propositions by giving the false impression that it is true that their logical conclusion is absurd. Ergo, I proffer that my theories are a good idea and are a valuable intellectual property, because, according to Abraham Lincoln, "You can't fool all of the people all of the time." Not even if you are the Department of State hospitals or CONREP! To disprove my past CONREP's proposition under their Dr. Douglas Porky, PhD, by showing his proposition is de facto, my brilliant theory's absurd contradictory, please just read my dossier, *A Three-Part Discussion Including Part Four*. And again, at this time, I proffer and profess again that we need all the formal, the official, and the authoritative opinions, opine, and also the wicked wit of all the cosmologists and astrophysicists we may be fortunate enough to afford and lure into my next hearing or trail in San Francisco's County Superior Court.

---

Where there is no sane and mitigating excuse for any of my civilian psychiatrists, and whereas I have these sane mitigating excuses, what California federal, state, or city political machine and forensic system should not rather lie double entendre above our California laws and queer me with reductio ad absurdum and homosexual rape due to (for me, horrible torturing side effects of) abject drugs like my Prolixin and my Haldol, rather than execute me for murder or send me to a place more prisonlike than Napa State Hospital?

Do I really need psychiatric care and treatment? If I must have psychiatric care and treatment, then which is best? Where the induced hallucinations are just our universe's future cosmological paradigms, if it is an argument or question of a psychotropic vs. a psychedelic, then my drug of choice is the powerful $C_{20} H_{25} N_3 O$ lysergic acid diethylamide, hands down, in 65–50 mg. then in 25 mg. doses once a month on Sundays. When I was among the lunatic avant-garde in the early 1970s, I used, real to me, TV hole or theoretical black hole paradigms in my stellar evolution before it was a putative and an acceptable academic thing to do, like it is in today's astrophysics. Way back then, in the 1970s, I theorized that large TV hole or theoretical black hole paradigms in my stellar evolution were in the center of spiral galaxies. I theorized they were probably key in the workings of mysterious quasars. And furthermore, back in the early 1970s, I posited that perhaps TV hole STS (space-time spheres) look like a theoretical black hole universe in my RCTVU. I had all those posits correct way back then, and I was an unheard of 4.0 at City College of San Francisco's community junior college in Stellar Evolution, and I occasionally took some small doses of LSD. The mystery is, why do I need a "double blind medication holiday" drug test to get at truths like unless the second (temporal and sufficient) wrong is to decide or determine who and what was wrong in the first place, two wrongs do not make an efficacious right. Ergo, it was wrong that there was the (discipline of psychiatry's) need for Atascadero State Hospital and Napa State Hospital to homosexually rape for the Department of Veterans Administration with any drug as long as it was not the powerful $C_{20} H_{25} N_3 O$, lysergic acid diethylamide, hands down, in 65–50 mg. then in 25 mg. doses once a month on Sundays!

Sometimes in court, the measure of a client is the measure of his or her attorney-at-law, and sometimes the measure of the client's attorney-at-law is the measure of the relationship between the client and her or his law firm. I need a winning attorney-at-law because the Department of State hospitals is sometimes extremely mean to me.

They have sometimes been party to and participated in my "cruel and unusual punishment." And moreover, they often prevaricate on my august issues in high places (courts) where they preponderate. Either as I say I care about myself and others, or as our Department of State hospitals have far too often said I am a danger to myself and others. So I need you as my attorney to stop them from assuming and avoiding or "begging the questions" before my court in my many, many forensic affairs, especially at my next PC 1606 or PC 1608 hearing or trial.

Please empathize. I am networking for mutual assistance. This informational gambit of information is also going to Bernard New England so you two may confer, focus, and be on the same pages, so to speak, regarding my case.

Thank you.

Respectfully submitted,

Mr. Dorian Gaylord Redus
Psychiatric patient

# A QUOTIDIAN QUASH: FROM MENTAL HYGIENE TO MENTAL HEALTH

Mr. Dorian Gaylord Redus
Napa State Hospital
Ward T-15
2100 Napa-Vallejo Hwy.
Napa, CA 94558-6234
United States of America
1(707)252-9988

Saturday, March 3, 2012

Dr. Steven W. Hawking
C/O Cambridge University
The Old School, Trinity Lane
Cambridge, CB2 1TN England

Re: a quotidian quash anywhere being a quotidian quash everywhere and cosmology.

*Legal Status:* I am a sixty-five-year-old light-skinned African American male under a PC 187, murder. On August 9, 1974, I stabbed a woman I had known for six years to death. In October 1975, I was found not guilty by reason of insanity and remanded to Atascadero State Hospital. I was transferred to Napa State Hospital in 1982, and I was released onto outpatient status in September 1988, under the supervision of San Francisco's CONREP. After leaving, I returned to Napa from June 1994 to May 2001. Atascadero and Napa state hospitals are in California. Finally, on October 1, 2009, I was again returned to Napa for keeping the angry voices I had been hearing for months a secret. And on May 5, 2010, my outpatient status was revoked. Now I want to win a sanity hearing or trial this year, 2012. And since I sent my public defender attorney-at-law my cosmological treatise, *A Three-Part Discussion*, on January 11 last year, she did not return even one of the weekly telephone calls that I placed to her until Thursday, December 1, last year. However, Friday, December 9, 2011, Cheryl H. Arkansas visited me here at Napa's visiting center.

To Whom It May Concern:

Thank you for receiving this letter. I am sixty-five years old. The bound twenty-page document is one half of my life's work. Only the last three pages are changed and new.

I think and I feel sending you my treatise, *A Three-Part Discussion*, in its unfinished form on Monday, October 20, 2011, helped me to make my (enclosed) finished treatise, *A Three-Part Discussion Including Part Four*. Again, thank you for all your past help—by receiving—time, and consideration. However, please do bless me, forevermore, with your insightful institutional evaluative response and reply on this my humble photocopied and bound work. It will be nice to have your reply.

In the 1990s, my mental health system called the part of this work that I had then my delusion about my universe. However, as my work was once summarily dismissed by some nontherapeutic medical doctors now, I am keeping hope alive that you will free me by stopping the preponderating forensic forces here from considering my work to be so bizarre as to be unworthy of consideration.

This is a last best hope—what American NCAA (college basketball) color television spectators call a game-winning Hail Mary three-point play. I can arrange a payment for any opinion given regarding my intellectual property, *A Three-Part Discussion Including Part Four*. Today, this is a tall order. As I write, it may be too tall. However, please send me your written reactions to my written work.

When you acknowledge your receipt of my work, *A Three-Part Discussion Including Part Four*, please know that I have no internet service here, and I am very technologically embarrassed in other ways. As a patient here, my only quash fighting option is my computer word processing in a program suite new to me on a 3.5" floppy disk and printing my work on it in a lab half a mile away in another building. However, this is a path that all staff here may accept, follow, and allow with their full approval and personal approbation. We have their support.

The world is a dangerous place. Not because
of the people who are evil, but because of the

people who don't do anything about it. (Albert Einstein in *The Week* for 11-25-11, page 23)

My own evaluative response regarding this, my new enclosed work, is "under a cloud" of quash, and all I think and feel is that a quotidian quash anywhere is a quotidian quash everywhere. Furthermore, I also think and feel that God is within us all, sometimes within one as one in their daily life, but I do not want the devil, even sometimes, in me as me in my daily life. So I obsecrate for our guidance.

My own conclusion is here the putative is just to be punitive, and our own dignity draining and destroying regressive "repressed difficulties and disagreements" afflict us all, severely worsening and aggravating Napa State Hospital's patients and staff alike. Only the last three pages of my manuscript are new or changed.

Again, thank you for receiving. Have a nice day.

Thank you.

Respectfully submitted,

Mr. Dorian Gaylord Redus
Psychiatric patient

A Three-Part Discussion
(1) Recession Velocity, (2) the RCTVU,
and (3) a Space-Time Sphere
(4) Including Part 4

Dorian Gaylord Redus

Compiled and Rewritten at
2100 Napa Vallejo Hwy.
Napa, CA 94558-6234
On Tuesday, January 17, 2012

# Eureka

## Cosmology

It is from numberless, diverse acts of courage and belief that human history is shaped. Each time a person stands up for an ideal, or acts to improve the lot of others, or strikes out against injustice, he [or she] sends forth a tiny ripple of hope.

—Robert F. Kennedy

# ABSTRACT

The subject in the first part, recession velocity, is the expansion and unfolding of space-time between and separating unmoving galaxies. This document's middle section, the RCTVU (relativistic color television universe), has two focuses or foci: (1) how to conceptualize our relativistic color television or its universe and (2) how to imagine a four-dimensional space-time. The third part of this precis is from a space-time sphere. If a naked singularity is an event horizon entered and exited, then the inside of a naked singularity may loom large, and thus an STS (space-time sphere) may be entered and imagined exited by picturing entering its event horizon and next exiting it through its center singularity womb of space-time, as "cosmic censorship" hides it all.

# RECESSION VELOCITY

Albert Einstein's special relativity equation for adding two velocities states unequivocally neither matter nor information may surpass the velocity of light. For almost one hundred years, observed facts have substantiated astronomer Edwin Hubble's original 1929 observations, proclaiming and proving the universe is a universe of galaxies that recede from one another at a recession velocity intrinsically based on their distance apart. Furthermore, no matter how arcane it may be, Dr. Einstein's work and Dr. Hubble's work may cause the unthinkable non sequitur, that both statements may not be true.

One light-year is the distance light travels in one year—nine trillion, 460 billion kilometers or five trillion, 880 billion miles. One parsec is a unit of astronomical distance equal to 3.26 light-years, and one megaparsec (Mpc) is a unit of astronomical distance equal to one million parsecs. Astronomer Edwin Hubble's law clearly states and elucidates that galaxies 4,001 megaparsecs or more distance apart produce a recession velocity greater than the speed or the velocity of light. Hubble's law is the recession velocity of a galaxy equals Hubble's constant (75 km/s/Mpc) times the distance usually given in Mpc between two galaxies. Or Hubble's law is the recession velocity of a galaxy divided by Hubble's constant (75 km/s/Mpc) equals the separating distance usually given in Mpc. However, we must remember that Albert Einstein's special relativity equation for adding two velocities states unequivocally neither matter nor information may surpass the velocity of light; therefore, special relativity's equation for adding two velocities is a speed limit of recession velocities and a size limit. Thus, it seems we may not link or concatenate megaparsecs of distance infinitely without being at a non sequitur or in error.

However, arcane Einstein's work and Hubble's work may be their work is not specious: seemingly true but false. My work above—the way I put them together—is specious, or seemingly true but false. Nevertheless, I am getting ahead of myself here. So what I am getting at will be gotten to in a moment.

Given the speed or velocity of light is 300,000 km/s, recall Hubble's law is the recession velocity of a galaxy equals Hubble's constant (75 km/s/Mpc) times the distance usually given in Mpc between two galaxies. Or Hubble's law is the recession velocity of a galaxy divided by Hubble's constant (75 km/s/Mpc) equals the separating distance usually given in Mpc.

So that

$$75_{(km/s/Mpc)} \times 4{,}000_{(Mpc)} = 300{,}000_{(km/s)}$$

Seventy-five kilometers per second per megaparsec times four thousand megaparsecs equals three hundred thousand kilometers per second or the speed or velocity of light.

Or

$$300{,}000_{(km/s)} / 75_{(km/s/Mpc)} = 4{,}000_{(Mpc)}$$

Three hundred thousand kilometers per second divided by seventy-five kilometers per second per megaparsec equals the distance four thousand megaparsecs.

And

$$4{,}000_{(Mpc)} = 4 \times 10^9_{(parsecs)} = 13.04 \times 10^9_{(Ly)}$$

Four thousand megaparsecs equals four billion parsecs equals thirteen point four billion light-years of age in time and diameter in size.

First, the following *words* of Dr. Mario Livio, an astronomer connected with the Hubble Telescope program, compared to my previous specious words make a paradoxical—*truth* opposed to common sense. When I found Dr. Livio's words on the expansion of the universe, especially between galaxies and clusters of galaxies, I had been looking for the answers they provided me for some thirty years. I believe that my previous words are a question, a test, and the answer follows:

>Second, I would like to clarify that it is only the scale of the universe at large, as expressed by the distances that separate galaxies and clusters of galaxies, that is expanding. The galaxies themselves are not increasing in size, and neither are the solar systems, individual stars, or humans; space is simply unfolding between them. Furthermore, a common misunderstanding is to think of the galaxies as if they are *moving* through some pre-existing space. This is not the case. Think of the dots on the surface of the [previously mentioned] balloon. Those dots are not moving at all on the surface (which is the only *space* that exists). Rather, *space itself* (the surface) is stretching, thus increasing the distances between galaxies. Finally on this point, the limit from special relativity that matter cannot move faster than light does not apply to the speeds at which galaxies are separated from each other by the stretching of space. As I explained above, the galaxies are not really moving, and there is no limit on the speed with which space can expand. As an aside, I should note that unlike in the case of moving cars or trains, the doppler effect used to determine the red-shift is also in this case a stretching of the wavelength due to the expansion of space itself. (Mario Livio, *The Accelerating Universe: Infinite*

*Expansion, the Cosmological Constant, and the Beauty of the Cosmos,* 49)

So the actual answer and my new point of view is this: what I thought and felt was just the recession velocity is also the everywhere increasing amount of stretching and unfolding of space that separates unmoving galaxies and unmoving clusters of galaxies in our big bang's Hubble bubble—the part of our universe photographed by the Hubble Telescope.

Let's calmly gawk at and ponder this relativistic speed limiting equation in two forms, prose and formula. When you add Velocity A plus Velocity B divided by one plus left hand parenthesis Velocity A multiplied by Velocity B divided by the speed of light squared right hand parenthesis, the sum of two relativistic velocities will never exceed the velocity of light.

$$Va + Vb = \frac{Va + Vb}{1 + \left(\frac{Va \times Vb}{C^2}\right)}$$

One may think of that special relativity equation for adding two velocities and the following special relativity equation for relative motion and time as two siblings. Albeit there are five of them. Let's calmly gawk at and ponder the following Lorentz transformation—the special relativity equation for describing relative motion and time in two forms, prose and formula. Time slows with velocity; the time in motion is proper, rest, time divided by the radicand, one minus left hand parenthesis the velocity of motion divided by the velocity of light, right hand parenthesis squared.

$$T = \frac{T_0}{\sqrt{1 - \left(\frac{V}{C}\right)^2}}$$

*Example*: Suppose that your friend is moving at 98 percent of the speed of light.

Then,

$$\frac{V}{C} = 0.98$$

So that

$$T = \frac{T_0}{\sqrt{1-(0.98)^2}} = 5T_0$$

Thus, a phenomenon that lasts for one second on a stationary clock is stretched out to 5 seconds on a clock moving at 98% of the speed of light. As measured by your fast-moving friend, a 60-second commercial on your TV will last for five minutes and the minute hand on your clock will take 5 hours to make a complete sweep. (William Kaufmann and Roger Freedman, *Universe*, 5th ed., 592–593)

# THE RCTVU

If you have ever wanted to read Dr. Brian Greene's book *The Elegant Universe* or wanted to read Dr. Stephen W. Hawking's three books—*A Brief History of Time, A Briefer History of Time,* and *The Nature of Space and Time*—then you may like perusing this, my essay, on the fabric of relative motion, mass, energy, length, space-time, and special relativity. In Brian Greene's book, relative motion may be perceived in perceptions of George's and Gracie's cell phone conversation and reception, and at the same time relative motion and "travel time effects" may be perceived in this, my essay, imagined on two color television screens.

Let's see this, first, from George's perspective. Imagine that every hour, on the hour, George recites into his cell phone, "It's twelve o'clock and all is well," "It's one o'clock and all is well," and so forth. Since from his perspective Gracie's clock runs slow, at first blush he thinks that Gracie will receive these messages prior to her clock's reaching the appointed hour. In this way, he concludes, Gracie will have to agree that hers is the slow clock. But then he rethinks it: "Since Gracie is receding from me, the signal I send to her by cell phone must travel even longer distances to reach her. Maybe this additional travel time compensates for the slowness of her clock." George's realization that there are competing effects—the slowness of Gracie's clock vs. the travel time of his signal—inspires him to sit down and quantitatively work out their combined effect. The result he finds is that the travel time effect more than compensates for the slowness of Gracie's clock. He comes to the surprising conclusion that Gracie will receive his signals proclaiming the passing of an hour on his clock after the appointed hour has passed on hers. In fact, since George is aware of Gracie's expertise in physics, he knows that she will take the signal's travel time into account when drawing conclusions about his clock based on his cell phone communications. A little more calculation quantitatively shows that even taking the travel time into account, Gracie's analysis of his signal will lead her to the conclusion that George's clock ticks more slowly than hers.

Exactly the same reasoning applies when we take Gracie's perspective, with her sending out hourly signals to George. At first the slowness of George's clock from her perspective leads her to think that he will receive her hourly messages prior to broadcasting his own. But when she takes into account the ever longer distances her signal must travel to catch George as he recedes into the darkness, she realizes that George will actually receive them after sending out his own. Once again, she realizes that even if George takes the travel time into account, he will conclude from Gracie's cell phone communications that her clock is running slower than his.

So long as neither George nor Gracie accelerates, their perspectives are on precisely equal footing. Even though it seems paradoxical, in this way they both realize that it is perfectly consistent for each to think the others clock is running slow.
(Brian Greene, *The Elegant Universe*, 45–46)

One should perceive the slow-motion clock and the "travel time effect's" peculiarity in space-time rates above as in the cell phone reception of George and Gracie in relative motion without acceleration as described and with exacting literalness. One should also perceive or see color television broadcasts, George should send to Gracie, and vice versa, broadcasts Gracie should send to George, in relatively slow motion because of the slow-motion clock and the "travel time effect's" peculiarity in space-time rates. These are both because of a governing speed limit, and that limit is the velocity of light, approximately 186,283 miles per second or about 670 million miles per hour. Therefore (this gets deep), one should also perceive and see in accordance with the physical influence of an equation: Einstein's relativity equation for space-time, as the "travel time effect" is also being considered. However, in spite of the fact that George and Gracie perceive each other's hourly messages—"It is twelve o'clock and all is well," "It is one o'clock and all is well," and so forth, after twelve o'clock and after one o'clock on their own clock—their own perception of their own time is proper and perceived as proper time is perceived here on earth.

Looking and calmly gawking at the following first four of the five beautiful special relativity equations for *mass*, *energy*, *length*, and *time*, we can (may) spell MELT and find the following: (1) the physics of (see note 2 page 300) *mass*'s increase with velocity can be imagined on a color television screen, viewed as a relativistic phenomenon, and (this gets deep) used to see, discover, that it is a manifestation or part of our relativistic color television universe. (2) The physics of *energy*'s equivalence with mass can be imagined on a color television screen, viewed as a relativistic phenomenon, and (this gets deep) used to see, discover, that it is a manifestation or part of our relativistic color television universe. (3) The physics of *length*'s decrease in the direction of motion with velocity can be imagined on a color television screen, viewed as a relativistic phenomenon, and (this gets deep) used to see, discover, that it is a manifestation or part of our relativistic color television universe. (4) The physics of *time*'s slowing with velocity can be imagined on a color television screen, viewed as a relativistic phenomenon, and (this gets deep) used to see, discover, that it is a manifestation or part of our relativistic color television universe. (5) Finally, that the adding of two velocities will never exceed the velocity of light can be imagined on a color television screen, viewed as a relativistic phenomenon, and (this gets deep) used to see, discover, that it is a manifestation or part of our relativistic color television universe. The ubiquitous motif, not ridiculous motif, relativistic color television universe theory and its slow-motion clock and its "travel time effect" may be used to describe all of this until we have the best way to describe our mental energy's relativistic existential universe. Moreover, I know, I think, and I feel these five equations of Albert Einstein are the symbols of chi in the United States of America. And furthermore, the relativistic color television universe is, in essence, the way to (this gets deep) discover American chi aborning in the twenty-first century because, for one thing, television was invented in the United States of America. See also number 10 in the bibliography and note 1 at the end of that section.

Try finding the relativistic color television and the relativistic color television universe in the following two excerpts from page 35 and page 36 respectively of Dr. Alan Guth's book *The Inflationary Universe: The Quest for a New Theory of Cosmic Origins*.

> This basic premise of special relativity, however, seems at first to be nonsensical. Suppose, for example, that I am at rest, and a light beam passes me. The speed would have to be c, the standard speed of light. Suppose, now, that I take off in a spaceship to chase the light beam at 2/3 of the speed of light. Common sense (or Newtonian physics) implies that I would then see the light pulse receding from me at only 1/3 of c. The premise of special relativity, however, holds that I would still measure the recession speed as c. No matter how hard I might try to catch a light beam, I will always see the beam recede at c. (p. 35)

Cogently, time slows for the fast-moving spaceship chasing the beam of light, so the beam of light always moves, is always seen, at c, the speed of light.

According to Einstein, the spaceship will appear to us to be shorter than the length that would be measured by observers inside the ship, although the width would be unchanged. The clocks on the ship would appear to us to be running slowly, and the clocks at the back of the ship would look like they were set to a later time than the clocks at the front of the ship, even though they would look synchronized to the crew. (Alan Guth, *The Inflationary Universe: The Quest for a New Theory of Cosmic Origins*, 36)

The phenomena of the preceding quotes above are all due to the relativistic, speed of light, television universe's limitation and constraint on movement, causing the speed of light to be a part cosmic television. See the curious and non-intuitive four dimensions defined in the motif above in the velocity of movement, and you may also imagine four-dimensional space-time.

The (late) Dr. Steven W. Hawking, Lucasian professor of mathematics at the University of Cambridge who, powered by his intelligence, was living on and on despite his terminal medical diagnosis, alleges that: "It is often helpful to think of four coordinates of an event as specifying its position in four-dimensional space-time." Where four-dimensional space-time includes length, width, height, and time inside a fifth dimension, I prefer to avoid Dr. Steven W. Hawking's "It is impossible to imagine a four-dimensional space." The preceding pair of quotes are from page 24 of Dr. Hawking's book *A Brief History of Time*. And in it, Hawking not only states categorically four-dimensional space is unimaginable, but he also states that it is relativistic. This is, of course, where relativistic means velocities great enough to approach the velocity of light. Likewise, one of the quirks of the world's scientific community (in general) is that it also maintains that a four-dimensional space-time continuum is relativistic. Where it is *not* impossible to imagine a four-dimensional space-time, if a space-time continuum is relativistic and four-dimensional, then it is part of the relativistic color television universe. Here, the term *relativistic color television universe* is my own (previously) covert concept. However, if it is developed overtly, it may become a term for the twenty-first century's scientific circles and communities at large. Please use it to gain insight into the esoteric realm of relativistic physics. My hope for this enterprise is that it will make the tool relativistic color television universe a user-friendly and honest intellectual edge for helping the scientific community to easily conceptualize—inter alia, among other things, inter alios, among other people—relativistic space-time.

To begin, visualize a seemingly two-dimensional color television as found in most American homes. Paint a mental picture on the screen, but then notice at the same time the two-dimensional screen is also three-dimensional—one height, two width, and three time (ignoring the dimension of depth, unless it is, in the first place, a 3D TV. One may then accept that electronic color televisions show space-time by imagining a clock or a Timex wristwatch as it accurately keeps space-time in your imagination. Furthermore, one may also accept that electronic color television pictures show space-time and some moving astral bodies and vehicles: planets, stars, galaxies, spaceships, and flying saucers. Then one may envision a flying saucer in a four-dimensional environment. Of course, the saucer will never actually leave the screen. On the other hand, one may imagine it traveling at relativistic speeds—receding from the earth, the solar system, and the galaxy. Finally, if we simply augment our cognizance with a second screen, a crucial element of space-time maybe comprehended. Also, remembering the George and Gracie cell phone excerpt, try this next suggestion sitting in front of your TV or computer tonight. Imagine seeing two color television screen images—one on the earth and the other in an extraterrestrial flying saucer out in space receding from the earth. Of the two imagined television screen images, one or both must have relative nonaccelerating motion to the other one and be approaching c, say 98 percent or 99.9997 percent of the speed of light. Next and equally important, they must both be imagined to be broadcasting an electronic color television signal to the other. George broadcasts cosmic TV to Gracie, and Gracie broadcasts cosmic TV to George.

Since the two broadcasting televisions one on earth and the other in a flying saucer are imagined to be moving relativistically to each other, one should also imagine that their rate of broadcast action on the receiving TV set screens is different, slower, coming from the flying saucer to earth and slower coming from earth to the flying saucer. The idea is to make the perceived paradoxical slow-motion clock's peculiarity in broadcast space-time rate in accord with the physical influence of an equation: Einstein's special relativity equation for space-time when in relative motion as the "travel time effect" is also being considered. Both screen's rate of received broadcast space-time should be imagined slower than a Timex wristwatch accurately keeping un-broadcast proper space-time on earth or in the flying saucer. Imagine seeing the earth's TV broadcasting to the flying saucer in relativistically slower motion, and imagine seeing the saucer's TV broadcasting to the earth in relativistically slower motion.

> These effects are summarized in relativity by saying that when someone watches an object recede away from them, that object will be seen to undergo mass dilation, length changes, and time dilation. (David Bodanis, *E=mc²: A Biography of the World's Most Famous Equation*, 82)

The TV set broadcast viewers on earth will see mass dilation, length changes, and time dilation in the flying saucer's broadcasting to them. The TV set broadcast viewers in the flying saucer will see mass dilation, length changes, and time dilation in the broadcasting to them from earth. This arcane relationship is always there and real. However, it is only obvious at most queer velocities or speeds like 98 percent or 99.9997 percent of the velocity or speed of light! Indeed, most readers will require more than an imagined proof. This will come when mankind acquires a real vehicle or phenomenon that is relativistic (see note 3 in this treatise). Then only a few readers will assume RCTVU: every man's relativity theory is erroneous and utterly ridiculous unrealistic craziness. Even now, some more sapient readers may think and feel this four-dimensional thought experiment is not a specious twenty-first century argument, and they may even understand that the all is not an electronic color television, but the all is a relativistic color television universe. To some with discerning gray matter, the following statement may be an ordinary theoretical fact. The idem here is that the rate of broadcast space-time is different, slower, and it is sometimes so through Einstein's special relativity equation for space-time. Thus, the motif is more powerful as evidence than a mere analogy. It is a mathematical, a relativistic, and a visual reality.

> "The medium is the message."
> —Marshall McLuhan

Scientific experiments have proved that Einstein's special relativity equation for space-time expresses exactly how space-time slows with velocity. Thus, the ubiquitous medium Einstein's equation influences is as (includes) the color TV's electronic message, and the ubiquitous medium Einstein's equation influences is as (includes) the relativistic color television universe. Intrinsically, it is cosmic color television universe with the speed of light limiting and constraining motion.

One, we saw televisions show space-time and "travel time effect." Two, we imagined a space-time difference. And three, I explained this is because relativistic space-time is different and not because of some kind of push button illusion. If one imagines seeing the space-time difference mentally on the two imagined TV screens, then, a fortiori, it is even more likely that one will see this relativistic phenomenon everywhere. Moreover, if we see how to imagine with the following MELT and Velocity A plus Velocity B special relativity equations of Dr. Albert Einstein, then a fortiori, it is even more likely that we will all one day see the RCTVU (relativistic color television universe). The continuum of space-time is where and when it is *not* impossible to imagine or see four-dimensional space-time.

Melting Albert Einstein's enamored equations down to prose and formula, we get:

M—mass increases with velocity. A mass in motion is proper, rest, mass divided by the radicand, one minus left hand parenthesis the velocity of motion divided by the velocity of light right hand parenthesis squared.

$$M = \frac{M_0}{\sqrt{1-\left(\frac{V}{C}\right)^2}}$$

E—energy has an equivalent mass. Energy equals mass times the velocity of light squared.

$$E = MC^2$$

L—length in motion shortens measured in the direction of motion. Length in motion, measured in the direction of motion, is the proper, rest, length multiplied by the radicand, one minus left hand parenthesis the velocity of motion divided by the velocity of light right hand parenthesis squared.

$$L = L_0 \times \sqrt{1-\left(\frac{V}{C}\right)^2}$$

T—time slows with velocity. The time in motion is proper, rest, time divided by the radicand, one minus left hand parenthesis the velocity of motion divided by the velocity of light right hand parenthesis squared.

$$T = \frac{T_0}{\sqrt{1-\left(\frac{V}{C}\right)^2}}$$

Va+Vb

Velocity A plus Velocity B

Because the velocity of light is "invariable," when you add Velocity A to Velocity B, divided by one plus left hand parenthesis Velocity A multiplied by Velocity B divided by the velocity of light squared right hand parenthesis, the sum of the two velocities will never exceed the velocity of light.

$$Va+Vb = \frac{Va+Vb}{1+\left(\frac{Va \times Vb}{C^2}\right)}$$

Our relativistic color television universe (RCTVU) must have the smaller scale.

According to string theory, the elementary ingredients of the universe are *not* point particles. Rather, they are tiny, one-dimensional filaments somewhat like infinitely thin rubber bands, vibrating to and fro. But don't let the name fool you: unlike an ordinary piece of string, which is itself composed of molecules and atoms, the strings of string theory are purported to lie deeply within the heart of matter. The theory proposes that *they* are ultra microscopic ingredients making up the particles out of which atoms themselves are made. The strings of string theory are so small—on average they are about as long as the

Planck length—that they *appear* point like even when examined with our most powerful equipment. (Brian Greene, *The Elegant Universe*, 136)

The smallness of the Planck's constant—which governs the strength of quantum effects—and the intrinsic weakness of the gravitational force team up to yield a result called the *Planck length*, which is small almost beyond imagination: a millionth of a billionth of a billionth of a billionth of a centimeter (ten to the minus thirty third centimeter)... If we were to magnify an atom to the size of the known universe, the Planck length would barely expand to the height of an average tree. (*The Elegant Universe*, 130)

Above the Planck length, our universe may just be a cosmic TV. Therefore, our universe may become a four-dimensional relativistic color television universe above the Planck length. And furthermore, our universe may also become a five-dimensional STS in a black hole's singularity under the Planck length, where it is dense enough. Clearly, we are all inside a Hubble bubble. Clearly we are all part of the RCTVU. And clearly, we all may have come from an STS that is open and expanding but contained by a ubiquitous five-dimensional quantum level. The following is an a priori description of a space-time sphere.

# A SPACE-TIME SPHERE

Where is all of this appropriate? Where there is understanding and tutelage for the proclaiming prose and the mathematics from Dr. Stephen W. Hawking's work *The Nature of Space and Time*.

> One could imagine that after being created, the black holes move far apart into regions without magnetic field. One could then treat each black hole separately as a black hole in asymptotically flat space. (Stephen Hawking, *The Nature of Space and Time*, 56–57).
>
> As in the case of pair creation of black holes, one can describe the spontaneous creation of an exponentially expanding universe. One joins the lower half of the Euclidean four-sphere to the upper half of the Lorentzian hyperboloid (fig. 5.7). Unlike the black hole pair creation, one couldn't say the de Sitter universe was created out of field energy in a preexisting space. Instead, it would quite literally be created out of nothing: not just out of the vacuum, but literally be created out of absolutely nothing at all, because there is nothing outside the universe. (*The Nature of Space and Time*, 85)

Therefore, can one create an inflationary period and the big bang out of nothing? Can one "describe the spontaneous creation of an exponentially expanding" cosmic TV universe out of nothing? And can one create an STS out of absolutely nothing?

If a naked singularity is an event horizon entered and exited, then the inside of a naked singularity may loom large. The following descriptive excerpts are on a black hole's way to make a big bang or space-time sphere. An STS, like a naked singularity, may be entered and imagined exited picturing, entering, its event horizon and next exiting it through its center singularity.

> In *The Life of the Cosmos*. Lee Smolin... posits that a process of self organization like that of biological evolution shapes the universe, as it develops and eventually reproduces through black holes, each of which may result in a new big bang and a new universe. (Lee Smolin, *The Life of the Cosmos*, back cover)
>
> If time ends, then there is literally nothing more to say. But what if it doesn't? Suppose that the singularity is avoided, and time goes on forever inside of a black hole. What then happens to the star that collapsed to form the black hole? As it is forever beyond the [event] horizon, we can never see what is going on there. But if time does not end, then there is something there, happening. The question is what?
>
> This is very like the question about what happened "before the big bang" in the event that quantum effects allow time to extend indefinitely into the past. There is indeed a very appealing answer to both of these questions, which is that each answers the other. A collapsing star forms a black hole, within which it is compressed to a dense state. The universe began in a similarly very dense state from which it expands. Is it possible that these are one and the same dense state? That is, is it possible that what is beyond the [event] horizon of a black hole is the beginning of another universe? (*The Life of the Cosmos*, 87–88)

Ersatz—substitute, artificial, or replacement. Something new is always a phantom, an error, or a queer until it is old hat! Is Lee Smolin's universe in a singularity a specious and ersatz connection? Should I believe it is when I have maintained the same connection? I trust the truth in his theory because I have maintained the same through my own heuristic old hat methods. Neil Alden Armstrong's 1969 moonwalk on the moon discussed in 2012. I came up with Dr. Lee Smolin's idea, and I called it a space-time sphere in 1972. However, Neil Alden Armstrong, the first human to walk on the moon, has probably not asserted his description of that 1969 moonwalk on the moon and felt as refuted and neglected as I felt describing my 1972 space-time sphere before reading Dr. Smolin's work *Life of the Cosmos*.

Intrinsically, my space-time sphere paradigm is our observable universe with its inflation and its big bang confined at a singularity's quantum level. And therefore, our singularity's function is to confine others inside relativistic frames of reference in other inner singularities, that each may hollow out, inflate, and form an independent space-time sphere for its own stellar evolution to connect its own intelligent life to its stars, in addition to confining our outside relativistic frame of reference outside the frames of reference inside other inner singularities. Our universe is, thus, that old hat big bang contained inside our STS. And therefore, the inside of our a priori STS is, by relativistic frames, of reference larger inside than outside.

The laws of physics break down at a singularity, and therefore, I posit, think, and feel the speed of light inside our space-time sphere is lesser than the speed of light outside our space-time sphere. The speeds of light inside and outside of our space-time sphere are not the same ineluctably tied together inside to outside. However, each separate inner and outer, both, speeds of light are the same related (tied) to just its own inside or outside frame of reference. The inside is, therefore, identical to its outside. If the respective velocities of light are related to just their own inner or outer frame of reference, then both of the velocities of light are identical. Again, the two, the inner and the outer speeds of light (are), may be ineluctably the same inside to outside (tied) to their own frames of reference. However, the two, both inner and outer speeds of light (are), may be somehow dynamically differential slower inside, making the inside larger inside than outside, if they are related (considered together) and not related to just their own inner or outer frame of reference.

In this paradigm a priori reality, the velocity of the propagating light inside the space-time sphere actually gives it its internal sizing. As in recession velocity, the first paradoxical part of *A Three-Part Discussion*, the space inside this paradigm a priori space-time sphere unfolds and stretches. Again, the velocity of light actually proportions the vast inner space, and the inside of this STS seemingly expands exponentially and forever, as if it were without an outside to the STS universe. Moreover, the STS womb of space-time—all space-time spheres—are most ineluctably immovable by their mass. However, if you do go over the speed or the velocity of light of or inside an STS and you are also part of an RCTVU (relativistic color television universe), then your movement moves—is a calculated action connected to—the original singularities of the space-time sphere. So it is still ad astra per aspera—to the stars through difficulties—unless we harness or control and direct the force of dark energy.

When I see velocities at and over the velocity of light, c + velocities, affect the stretching and unfolding of our exponentially expanding universe, I like to also realize that according to NASA's Wilkinson Microwave Anisotropic Probe (the WMAP satellite) launched in 2001, the universe is 23.3 percent dark matter, 72.1 percent dark energy, and the part of the universe that I base my work on—ordinary matter—is only 4.6 percent of my universe.

# ADDENDUM

This treatise's last two parts are either a growing cancerous delusion or a discovery of merit. If they are a delusion, then I can attack them with the appropriate erudition, as I have done in my treatise's first of its three parts, "Recession Velocity."

As for me, I am currently encouraged that Dr. Steven W. Hawking's cogent words, "It is impossible to imagine a four-dimensional space," from his 1988 book *A Brief History of Time*, are words on a subject Dr. Hawking "almost" completely omits from his more recent 2005 book, *A Briefer History of Time*. Perhaps the impossible was too difficult for too many of his readers to comprehend, or perhaps he was mistaken.

Moreover, in his 1997 book *The Inflationary Universe: The Quest for a New Theory of Cosmic Origins* author Dr. Alan H. Guth writes, on page 38, that:

> You must imagine a four-dimensional Euclidean space, and then imagine a sphere in the four-dimensional space. The three-dimensional surface of the sphere is precisely the geometry of Einstein's cosmology. (If you have difficulty visualizing a sphere in four Euclidean dimensions, rest assured that you have a lot of company, including the author)

Adducing further! Eureka! In my second reading of *A Briefer History of Time*, I found the following two sentences on page 141 of its conclusion:

> When we combine quantum mechanics with general relativity, there seems to be a new

possibility that did not arise before: that space and time together might form a finite, *four-dimensional space* (italics added) without singularities or boundaries, like the surface of the earth but with more dimensions. It seems that this idea could explain many of the observed features of the universe, such as its large-scale uniformity and also the smaller-scale departures from homogeneity, including galaxies, stars, and even human beings.

# BIBLIOGRAPHY

Bodanis, David. *E=mc²: A Biography of the World's Most Famous Equation.* New York: Walker, 2000.

Greene, Brian. *The Elegant Universe.* New York: W.W. Norton and Company, 1999.

Guth, Alan. *The Inflationary Universe: The Quest for a New Theory of Cosmic Origins.* Reading, Massachusetts: Perseus Books, 1997.

Hawking, Stephen. *A Brief History of Time.* New York: Bantam Books, 1988.

Hawking, Stephen. *A Briefer History of Time.* New York: Bantam Books, 2005.

Hawking, Stephen. *The Nature of Space and Time.* Princeton, New Jersey: Princeton University Press, 1996.

Kaufmann, William and Roger Freedman. *Universe.* 5th ed. New York: W. H. Freeman and Company, 1999.

Livio, Mario. *The Accelerating Universe: Infinite Expansion, the Cosmological Constant, and the Beauty of the Cosmos.* New York: John Wiley & Sons, Inc., 2000.

Primack, Joel R. and Nancy Ellen Abrams. *The View from the Center of the Universe: Discovering Our Extraordinary Place in the Cosmos.* New York: Riverhead Books, a division of Penguin Group (USA), 2006.

Ratti, Oscar, and Adele Westbrook. *Aikido and the Dynamic Sphere.* Boston, Massachusetts: Tuttle, 2001.

Smolin, Lee. *The Life of the Cosmos.* New York: Oxford University Press, 1997.

## Pamphlet

The special relativity equations are from James E. Bradner and Tamar Y. Susskind's *Theories of Relativity* (1995, West Crescent Avenue, Anaheim, California 92801. Litton Instructional Materials, Inc., a division of Litton Industries).

## Note 1

The first two successful television devices were Russian-born American physicist Vladimir Kosma Zworykin's 1923 iconoscope and American radio engineer Philo Taylor Farnsworth's image dissector tube.

## Note 2

Suppose a super space shuttle is blasting along very close to the speed of light. Under normal circumstances, when that shuttle is going slowly, the fuel energy that's pumped into the engines would just raise its speed. But things are different when the shuttle is right at the very edge of the speed of light. It can't go much faster.

Think of frat boys jammed into a phone booth, their faces squashed hard against the glass walls. Think of a parade balloon, with an air hose pumping into it that can't be turned off. The whole balloon starts swelling, far beyond any size for which it was intended. The same thing would happen to the shuttle. The engines are roaring with energy, but can't raise the shuttle's speed, for nothing goes faster than light. But the energy can't just disappear, either.

As a result, the energy being pumped in gets "squeezed" into becoming mass. Viewed from outside, the solid mass of the shuttle starts to grow. There's only a bit of swelling at first, but as you keep on pouring in energy, the mass will keep on increasing. The shuttle will keep on swelling.

It sounds preposterous, but there's evidence to prove it. If you start to speed up small protons, which have one "unit" of mass when they're standing still, at first they simply move faster and faster, as you'd expect. But then, when they get close to the speed of light, an observer really will see the protons begin to change. It's a regular event at the accelerators outside of Chicago, and at CERN...near Geneva, and everywhere else physicists work. The protons first "swell" to become two units of mass—twice as much as they were at the start—then three units, then on and on, as the power continues to be pumped in. At speeds of 99.9997 percent of "c," the protons end up 430 times bigger than their original size.

What's happening is that energy that's pumped into the protons or into our imagined shuttle has to turn into extra mass. Just as the equation states: that "E" can become "m," and "m" can become "E" 1. (David Bodanis, *E=mc2: A Biography of the World's Most Famous Equation*, 51–52)

## Note 3

People haven't traveled like this yet because our fastest rockets move far more slowly than light, but nature does this sort of experiment with elementary particles all the time. Unstable elementary particles called muons are the main component of the cosmic rays that reach low altitudes where most people live, but they would have decayed high in the atmosphere if their lifetimes were not greatly lengthened by this [time dilation] relativistic effect. These muons, which have a half-life of only 1.52 microseconds ($1.52 \times 10^{-6}$ s), live much longer when they move at nearly the speed of light, exactly as predicted by relativity. (William Kaufmann and Roger Freedman, *Universe*, 5th ed., 340–341)

# LAST WORD

Finally, the following is not holography to me. The following describes something analogous to a TV screen as is found in most American homes, or the following describes something analogous to a PC screen:

> One idea that has drawn a lot of attention follows from a startling theoretical discovery made by Stephen Hawking at the University of Cambridge, UK, and others in the 1970s: quantum effects in the space around a black hole cause it to emit radiation [Hawking radiation] as if it were hot, even though black holes are supposed to swallow mass and energy, not spit it out.
>
> Furthermore, after decades of analysis and generalization of this argument, many physicists now believe that it applies to any three-dimensional volume, from black holes to empty space: the volume's entire information content can be encoded in its two-dimensional surface. Or to put it another way, the ultimate unified theory of everything should describe our apparently solid three-dimensional world in terms of a lower-dimensional reality. Our Universe would emerge from the theory like a three-dimensional optical image from a two-dimensional hologram. [Here, I feel "hologram" is a misnomer for my relativistic color television universe at least outside an STS.] (M. Mitchell Waldrop, *Nature*, vol. 471, March 17, 2011, p. 288)

According to Sean Carroll:

> An ordinary hologram displays what appears to be a three-dimensional image by scattering light off of a special two-dimensional surface. The holographic principle says that the universe is like that, on a fundamental level [between the outside and the inside of an STS and on all objects with two or more dimensions inside the STS]: Everything you think is happening in three-dimensional space is secretly encoded in a two-dimensional surface's worth of information. The three-dimensional space in which we live and breath could (again, in principle) be reconstructed from [or on to] a much more compact description [on a quantum computer's monitor or a relativistic color television set's TV screen.] (Sean Carroll, *From Eternity to Here: The Quest for the Ultimate Theory of Time*, 280–281)

# LOOKING THE LOOKER

Being an unusually speculative speculator and an individual seeking reality, conjecturally, a space-time sphere probably looks like a TV hole. A TV hole probably makes our observable universe look just like a black hole universe, and our STS is probably just one of many STSs in our relativistic color television universe.

Although, all four of the following equations all use c or the speed or the velocity of light, and in them the $h=1.05457...\times 10^{-34}$ Joule seconds, it is a tall order to say, therefore, they are altogether based on c and evidencing QED (that which was to be demonstrated) my RCTVU theory. However, that is my comprehensible assumption or arrangement for the four separate equations as a group. Thus, the tall order of RCTVU is really to "Let there be light."

The Planck time, and the corresponding *Planck length* $l_p = ct_p$, are often regarded as providing a kind of 'minimum' space-time measure (or 'quantum' of time and space, respectively), according to common ideas about quantum gravity:

$$t_p = \sqrt{\frac{Gh}{c^5}} \approx 5.4 \times 10^{-44} s, l_p = \sqrt{\frac{Gh}{c^3}} \approx 1.6 \times 10^{-35} m.$$

By use of these Planck units and also the *Planck mass* $m_p$ and Planck energy $E_p$ given by

$$m_p = \sqrt{\frac{hc}{G}} \approx 2.1 \times 10^{-5} g, E_p = \sqrt{\frac{hc^5}{G}} \approx 2.0 \times 10^9 J,$$

> which are naturally determined (though completely impractical) units, one can express many other basic constants of nature simply as pure (dimensionless) numbers (and these numbers may enlighten us as to RCTVU and STSs at the quantum level).

The indented above is from page 163 of *Cycles of Time: An Extraordinary New View of the Universe* by Roger Penrose, professor of mathematics at the University of Oxford.

My discussion continues below in the work of theoretical physicist Sean Carroll, PhD:

> In Chapter Eight we contemplated an irreversible game of billiards: a conventional billiards table, where the balls moved forever without losing any energy through friction, except that, whenever a ball hit a particular one of the walls of the table, it came perfectly to rest and stayed there forever... But the dynamics are irreversible: Given any one ball stuck to the special wall, we have no way of knowing how long it's been there. And the entropy of this system flouts the Second Law with impunity; gradually, as more balls [RCTVU] get stuck, the system takes up a smaller and smaller portion of the space of states, [this is inside of the STS] and the entropy decreases with out the intervention from the outside world. (p. 342)

> So if we wish to explain the arrow of time… [QED]… It might seem that this setup gets it backward—it predicts that entropy goes down, rather than up… In other words, such observers would call the high-entropy end of time the "future," and the low entropy end "the past," even though the fundamental laws of physics in this world would only precisely reconstruct the past from the future, and not vice versa. [This is outside of the STS.] (Page 343)
>
> The universe [outside of an STS], for whatever reason, finds itself in a randomly chosen high-entropy state, which looks like empty de Sitter space. Now our postulated irreversible laws of physics act on that state to decrease the entropy. The result—if all this is to have any chance of working out—should be the history of our actual universe, just reversed in time compared to how we traditionally think about it. (Page 343)
>
> An explicit model of such a bouncing cosmology [inside to outside] was proposed by Anthony Aguirre and Steven Gratton in 2003. They based their construction on inflation and showed that by clever cutting and pasting we could take an inflationary universe that was expanding forward in time and glue it at the beginning to an [outside of the STS] inflationary universe expanding backward in time, to obtain a smooth bounce. (Sean Carroll, *From Eternity to Here: The Quest for the Ultimate Theory of Time*, p. 353)

---

Now my own suspicion is that the universe is not only queerer than we suppose, but queerer than we can suppose.
—J. B. S. Haldane

Irrefutably, now my own suspicion is that our theoretical universe is not only more relativistic than we suppose, it is also more relativistic than we can suppose.

The amassment and the essence that is my STS's big bang and its RCTVU's time, space, matter, and energy expand out to an external infinity like the pool balls on the sides of Sean Carrol's pool table seen as a simile/analogy—*in saecula saeculorum* for ever and ever, fulfilling the past hypothesis that the beginning has a low entropy requirement.

Mr. Dorian Gaylord Redus
Napa State Hospital
Ward T-15
2100 Napa-Vallejo Hwy.
Napa, CA 94558-6234
1(707)252-9988

Wednesday, April 11, 2012

Department of Veterans Affairs
Oakland Regional Office
Oakland Federal Building
1301 Clay Street
North Tower Twelfth Floor
Oakland, CA 94612
1(800)827-1000 Ex 110

Re: REDUS, DORIAN GAYLORD DOB: 05/19/1946

*Subject:* Redus, Dorian Gaylord and his current medication history. Just like, I cannot know how tolerated and efficacious my new Friday, January 27, 2012, Consta injections regimen is until I have waited from January 27 to Palm Sunday, April Fool's Day 2012. I will (and this is psychiatry's crime) never know one thing—*how sane or insane am I off my Consta injections?*

*Legal Status:* I am a sixty-five-year-old light-skinned African American male under a PC 187, murder. On August 9, 1974, I stabbed a woman I had known for six years to death. In October 1975, I was found not guilty by reason of insanity and remanded to Atascadero State Hospital. I was transferred to Napa State Hospital in 1982, and I was released onto outpatient status in September 1988, under the supervision of San Francisco's CONREP. After leaving, I returned to Napa from June 1994 to May 2001. Atascadero and Napa state hospitals are in California. Finally, on October 1, 2009, I was again returned to Napa for keeping the angry voices I had been hearing for months a secret. And on May 5, 2010, my outpatient status was revoked. Now I want to win a sanity hearing or trial this year, 2012. And since I sent my public defender attorney-at-law my cosmological treatise, *A Three-Part Discussion*, on January 11 last year, she did not return even one of the weekly telephone calls that I placed to her until Thursday, December 1, last year. However, Friday, December 9, 2011, Cheryl H. Arkansas visited me here at Napa's visiting center.

To Whom It May Concern:

Thank you for receiving this letter. Did you get my other manuscript? I sent it on November 7, 2011. Its parts one and two are the same as in this manuscript, which also has its new part three.

When I corrected about one-third of the written and preponderating misstatements in some court documents from California's Department of State hospitals to some of my San Francisco judges, they still continue to include the misleading misstatements. Videlicet, they called my late VA psychiatrist, Dr. Donald Montana, MD, who was the Veterans 1970s Chief of Mental Hygiene in San Francisco, even though he malpracticed as a Veterans Administration doctor—my private psychiatrist.

When I had reported that he had told me (once or twice) in his VA office to "get a gun," they wrote at a preponderating "intellectually dishonest" subterfuge or devise on a Department of State hospital's legal court document: "He [Dorian Redus] had informed his psychiatrist at the time that his girlfriend was having an affair. He heard his psychiatrist respond, 'Get a gun and shoot her.'" My victim in my 1974 homicide was Ms. Edna Ella Robenson.

I need two doctors who my California judge will listen to, to do three legal things: one, read about two hundred pages of my writing at $5 a page; two, see me here at Napa State Hospital's official visiting center to discuss my case with me and evaluate me; and three, I need them to legally and officially advise my court in San Francisco (I hope of my sanity) by also working with my attorney-at-law.

The preponderating oxymoronic lies at my court hearings and trials during the past forty years of litigation which are due to my cosmology theories, due to my homicide thirty-eight years ago. And moreover, the lies due to my hundreds of consensual homosexual rapes approximately thirty years ago are all unjustly an overwhelming juggernaut, over-mind, that lies (double entendre) just above our California laws. Videlicet, they recently wrote, "There is no known history of emotional or sexual abuse" on page 8 of a May 16, 2011, letter to the Honorable Master Calendar Judge San Francisco's County Superior Court by Dr. Eric Florida, MD, my current psychiatrist here at Napa State Hospital.

I have family here who have forgiven me. Through my essential bound photocopied manuscript of 222 pages, *A Quotidian Quash: From Mental Hygiene to Mental Health 1969–2012; Part 1, Part 2, and Part 3*, I have forgiven myself and garnered a great desire to get a life outside of California's new Department of State hospitals.

For me, and very much so, I would like a double-blind medication holiday from my current psychotropic medication as it has been and is currently indicated. Do I even have such a lie detector available to me such that something other than my currently (considered) insane "cosmic TV" mind adjudicates the adversarial mistakes and lies that work against me or work against my side being paid accordingly as a sane "cosmic TV mind?" Here and now at this state hospital where my psychiatrist's insanities are a perfidious breach of trust, there are sickening information relationships between my psychiatrist's motives and my motives. However, we really want the same thing—money.

> To rule out a problem from [those] information relationships, in the secure treatment area of this state hospital, a "medication holiday" was [yes] indicated, but I have been medicated here on Napa's Ward T-15, its best open ward, since September 15, 2010. And my recently requested "medication holiday" from my low dose of psychotropic medication for schizophrenia [to further what was started by my previous psychiatrist, Dr. Hameed Nebraska, MD, on my previous ward here Ward T-14] was denied on January 18,

2012, in utter disregard of my excellent behavior and recent election to Ward T-15's ward government as its vice president. Whereupon, I said, "If it is inevitable, then I want the medication change to Consta injections as soon as possible. [I was thinking, *When rape is inevitable, sit back and enjoy it.*] And then my psychiatrist, social worker, and psychologist all laughed as I laughed. (This is from my Monday, January 30, 2012, letter to two attorneys-at-law)

A posteriori, I feel contemptuousness in my thoughts on my basic civilian psychiatrists, and I think they have all grievously handicapped me. At life, nonfiction is in my case, stranger than science fiction. My lysergic acid use in the early 1970s was not the cause and is probably not responsible for my August 9, 1974, PC 187, murder. Here and now, my self or my ego are not sacrosanct because of my incarceration. It is slave to the psychiatric bombardment of my mammalian brain's limbic system, which is responsible for my subjective joys, my subjective pleasures, and my ego is slave to a contemptuous feeling or evil construct of my psychiatrists—my mental illness. This is all because I have always felt like telling my psychiatrists everything they wanted me to share with them. Whatever, those with the sufficient adequacy may also feel my desirable coattails are a therapist's to grab on to and follow as if I am one with leadership qualities. Way back in 1972, in the good old days, as my VA mental hygiene became a sequence of nasty California Department of State hospitals situations that undercut my mental health, there were no psychotropic medications for me from the Department of Veterans Affairs Chief of Mental Hygiene, the late great Dr. Donald Montana, MD. The mystery is, why do I currently need a "double-blind medication holiday" drug test to get at truths like unless the second (temporal and sufficient) wrong is to decide or determine who and what was wrong in the first place (see if it was not me and was Dr. Montana), two wrongs do not make an efficacious right. Ergo, it was wrong (in the early 1980s) that there was the (discipline of psychiatry's) need for Atascadero State Hospital and Napa State Hospital to homosexually rape for the Department of Veterans Administration with any drug as long as it was not the powerful $C_{20}H_{25}N_3O$, lysergic acid diethylamide, hands down, in 65–50 mg. then in 25 mg. doses once a month on Sundays!

When I trusted my first long-term VA psychiatrist, instead of being trustworthy, he begged the question of his personal responsibility for any of my psychiatric problems. He avoided his guilt, and he publicly assumed his innocence, transferring the probable cause and the guilt for my socioeconomic problems, socio-psychiatric problems, and other almost ballistic criminal self-defense problems from him to me. And furthermore, to me, he also feigned all of his respectability and was publicly different and unjust to me for our more than five years of one-on-one VA therapy. He was a sophisticated chief of psychiatry for the VA, and I was his trusting, naive, junior college patient. It was counterintuitive to the typical norm, but he was harmful, and I was helpful. After my murder and during its subsequent continuing decades of litigation, he, my original long-term treating psychiatrist, just left me holding the bag. And moreover, he never ever appeared for or against me in any court of law. Dr. Donald Montana, MD, was an abject (contemptible, despicable, probably miserable, and wretched) psychiatrist who I found to be grossly negligent, inadequate, unable to help me, and a very dangerous man whom I dare say here and now still makes my m-word malpractice as his nerve and sinew still serve him even though he long ago passed away. Both Edna and Donald had an immaculate way of gawking at me and not saying what was really on their minds.

## Moral Madness and Temporary Insanity

A fortiori, just like in the way your human brain's cortex extracts its rules of language, your human brain's orbital frontal cortex extracts its morals or rules of social interaction from the time you were born. In 1966, I made a judgment call that became a calling for me to think long and hard. I weighted and balanced "free love" and my neologism "cosmological TV," which became my RCTVU (relativistic color television universe) theory for our twenty-first *century*. And moreover, when we follow our moral code, our brain's limbic "endogenous reward system" makes us feel better. Conversely, we feel negative shame, guilt, and just plain worse when we fail to live up to our moral code. If "free love" and my neologism "cosmological TV," which became my RCTVU (relativistic color television universe) theory, are the only two foci or choices, then only one "cosmological TV" is useful to study psychiatry in the twenty-first *century* in America. And thus, it may also be used to study not only psychiatry but all of mankind as an evolving group with the selective advantage "cosmological TV." So we should use our influence to choose cosmological TV over free love.

Again, in 1966, I was a gentleman. I had no sexual relationships whatsoever with any of the hippies I saw or met in San Francisco where I lived. And to this day, that was a good but unheard-of coping mechanism. I sublimated my raging hormones from some first sexual intercourse at seventeen to my first (asexual) LSD at the age of twenty. Moreover, again, my lysergic acid use in the early 1970s was not the cause and is probably not even responsible for my August 9, 1974, PC 187, murder. The personally envied visiting City College of San Francisco lecturers, involved in CCSF's televised summer of 1972, *Stellar Evolution: Man's Descent from the Stars* lectures (that kept me interested in womankind), were Professor Duckworth's avant-garde astronomical group, disclaiming my spelling: Mr. Ray Bradbury, Dr. Geoffrey Burbidge, Dr. Edwin E. Salpeter, Dr. P. J. E. Peebles, Professor J. W. Schof, Dr. Sherwood Washburn, Dr. Melvin Calvin, Dr. Philip Morrison, Dr. Freeman Dyson, and Dr. Bernard M. Oliver of HP Co. I was connected to life; women; Ms. Edna Ella Robenson, my offense victim; human evolution; and to cosmology at the time of my August 9, 1974 PC 187, murder. They were my source of my sanity.

## My Temporary Insanity Caused my Murder

As Dr. Donald and I ignored her violence, Edna's violence created in me two jeopardizing psychoses. When Dr. Donald ignored my many disadvantageous disagreements with him and

Edna, the disagreements resurfaced in a general psychotic paranoia and a specific psychotic delusion. My suspicions and my disagreements caused paranoid conjecture, and my paranoid conjecture (due to my real jeopardy) did produce a delusion that people do not die. Those psychoses became the conjectures: I may only get the much-needed help I need if I take Edna's life, and that if I don't take her life soon, she will surely kill me first. Therefore, in clinical desperation, as a last resort, I temporarily chose the former—to take Edna's life, like I had learned in the United States of America's US Army. In conclusion, all of this information has been on the tip of my tongue and occasionally coming out of my mouth in words for going on forty years. However, all of this information, like the Department of Veterans Affairs late Dr. Donald Montana, has never come to court and been believed! (This is from my Monday, May 23, 2011, letter to the Honorable Judge Wyoming, San Francisco County Superior Court)

I said all of that to say this: I am finding myself sane (for decades), and I am finding, if it is about me, then it is either good or an *un*just construct of one or more of my many psychiatrists. I commend to you, my reader, whatever is wrong with your present picture of me, that is the half of it. The other half is, do you find it necessary to continue my (*un*bloody) crucifixion as described in my document, *A Quotidian Quash: From Mental Hygiene to Mental Health 1969–2012; Part One, Part Two, and Part Three*? Or do you find me, sans suspiciousness or conjecture, SANITY RESTORED already?

To me, Napa State Hospital is *un*justly just another Baby Babylonia. Furthermore, as a reluctant Babylonian here at this state hospital, there is one conceptually different but united state hospital for each individual patient here like me.

*Psilocybin for depression:* "Magic" mushrooms could have the capacity not only to blow users' minds but also to heal them. British neuroscientists injected volunteers with the chemical psilocybin, the psychedelic ingredient in hallucinogenic mushrooms, while scanning their brains. Since psilocybin mushrooms "are thought of as 'mind expanding' drugs," the scientists expected to see marked increases in brain activity, Imperial College London professor David Nutt tells *Nature News*. But to their surprise, activity decreased—especially in the parts of the brain that ground us in reality and govern our sense of self. Those regions—the medical prefrontal cortex and the posterior cingulate cortex—tend to be hyperactive in people with depression. The findings suggest that psilocybin's ability to give recreational users dream-like, out-of-body experiences could also help depressed patients break free of the "particularly restrictive state of mind" that forces them into loops of negative thinking, says study co-author Robin Carhart-Harris. The effects could also be long-lasting; previous research has shown that a single high dose of psilocybin can improve the mood of recipients for more than a year. (From *The Week* for February 10, 2012, page 19)

If and only if I must take a psychotropic or psychedelic medication rather than injections like Consta injections to help me to break free of psychiatry's demonic adversarial constructs, turn me on while scanning my brain, and diagnose me figuring stuff and things out by what lights up while I think of my "cosmological TV" and mankind in the *twenty-first century,* I feel that that will help my situational self or ego.

Please know my psychiatrist's past diagnostic constructs have been queer. However, I am heterosexual, still congenially, and still consensually funny, but my chronic apprehension/fear makes all my stress funnier and funnier. I cannot tell if I am inside and on my psychiatrist's super highway to safety or if I am inside and on another dicey and dangerous detour to *rape*, literally, or (heaven forbid) a situational one until it is homosexual. Fortunately, the prescribing "ignoramus" that is my current Consta prescribing nemesis cannot say it is no skin off his back if I copyright and publish. So I like readers that, like me, like to read my work, and when I write my work and read my work, it helps me to be writing for the American people.

On DVD disk no. 3's lecture no. 13 of the Great Courses Series, *Understanding the Brain* by Jeanette Norden of the Vanderbilt School of Medicine, I learned "cones" in the eye provide the data that allows the brain to construct or to "see" color. To elucidate my learning further, the eye does not see blue. The brain constructs the blue we see as blue. I also learned the following regarding the most putative problem that caused my most recent return to Napa State Hospital for the immediate eight months before my October 1, 2009, return to Napa State Hospital. My signs and symptoms (after my more than three years of intense sex offender therapy groups) were my constructed: auditory hallucinations or "angry voices" due to some therapists' daffy and dicey therapies for me. Like they wanted me convicted of molestation so bad. I constructed my therapist's angry voices because of my high average-superior range, October 7, 2009, Wechsler Abbreviated Scale of Intelligence (WASI) full-scale IQ of 124, 95$^{th}$ percentile, 95 percent chance actual estimated full-scale IQ is between 119–128, superior range.

> There are no blue cones that respond to short-wavelength light in the fovea [where the cones of the eye are]… So how is it that you can look directly at me (You can look directly at me.) and see my blouse is blue? How can that happen? Well! It happens because this is a construct created by your brain, and how color vision takes place is that the brain interprets the relative firing of the different kinds of cones and then creates what we subjectively experience as a particu-

lar color. So the brain is going to say, when you look directly at me with your fovea, that there are no firing of blue cones [She was wearing a blue blouse in the DVD video.], and the relative firing of the other types of cones is such that the brain creates and knows what's coming in "creates" a subjective experience that we say is blue. And in fact, this is how all color vision takes place. All color vision is the consequence of the brain's interpretation of the relative firing patterns of the different types of cones. Now—color is what we experience privately and subjectively.

When the mind constructs *un*real voices, this is a pathology and a disadvantage: 2+2=5. The answer is wrong. When the mind constructs the color blue, this is an evolutionary advantage: 2+2=4. The answer is right. As I hoped, it was 2+2=4. For forty years, I constructed and "fabricated" my cosmological RCTVU (relativistic color television universe) theory. Ipso facto constructing, when I hoped it was blue because of psychiatry, it was a gray area for forty years because psychiatry said it was a gray area or my pathological construct. I knew it was natural/right/objective, so I was driven to fabricate and construct my construct—"cosmological TV" theory.

Fortunately or *un*fortunately, since 2005, "out of the blue," my many therapists' minds also constructed their fabricated child molestation case against me. Fortunately I then knew my cosmological therapy was greater than all of my many therapist's molest, conjectures, put together. Unfortunately, my therapist's intransigent resolve to convict me of psychedelic/child molestation has lasted for decades and caused *un*real angry voices which caused me to return to this state hospital for the criminally and mentally insane where I now live on and on as an incarcerated client of yours.

My problem is, are my therapists' choices, constructs, and fabrications functional or dysfunctional? My relativistic "cosmological TV" is functional. Moreover, when I am innocent of psychedelic/child molestation, investigating me for it, treating me for it, and trying to convict me of it is a dysfunctional and bad choice that makes my therapists' choice to do so their child molestation. My past therapists are molesting all the children who could have come in contact with me! It is crazy! After seven years of my therapists' speculating groups and inquiry, the only speculation that is panning out is my speculation on my other cosmological theory, STS (space-time sphere). I am in therapy for an inappropriate loss of life, but worse, I am here for saying 2+2=4. And they are keeping me here by figuring one plus four is not five as I posit it is in my STS theory!

Where I fear my therapy is going next. On DVD disk no. 3's lecture no. 15 of the Great Courses series, *Understanding the Brain* by Jeanette Norden of the Vanderbilt School of Medicine, I learned the following about some people with an intractable pain problem that medication could no longer help:

> When they did prefrontal lobotomies or leukotomies, it seemed as though these people weren't in pain anymore, and then lo and behold, a physician asked the right question, he asked the individual "Do you still feel the pain?" and the person said "I feel it just like I always did. I just don't care anymore." The emotional component of the painful experience had been dissociated from the feeling of the pain itself, because, the emotional elaboration—which is that response—is a cortical function, and the front part of the cortex had been cutoff from the rest of the brain [in these people with intractable pain]. Cortex: Outer "bark" or mantle of the two cerebral hemispheres [of the brain]. Lobotomy: Removal of a cortical lobe.

*The rapist* was my *therapist*. Check the spelling! Now that it is the *twenty-first Century* and your evil doctor, Dr. Donald Montana, MD, also of Dartmouth University, is long gone, is the Department of Veterans Affairs still illegal to the bone? And do I still want to be military?

 US military researchers have had great success using "transcranial direct current stimulation" (tDCS)—in which they hook you up to what's essentially a 9-volt battery and let the current flow through your brain. After a few years of lab testing, they've found that tDCS can more than double the rate at which people learn a wide range of tasks, such as object recognition, math skills, and marksmanship.

 When a nice neuroscientist named Michael Weisend put the electrodes on me, what defined the experience was not feeling smarter or learning faster: The thing that made the earth drop out from under my feet was that for the first time in my life, everything in my head finally shut up.

 The experiment I underwent was accelerated marksmanship training, using a training simulation that the military uses. I spent a few hours learning how to shoot a modified M4 close-range assault rifle, first without tDCS and then with. Without it I was terrible, and when you're terrible at something, all you can do is obsess about how terrible you are. And how much you want to stop doing the thing you are terrible at.

 Then this happened: The 20 minutes I spent hitting targets while electricity coursed through my brain were far from transcendent. I only remember feeling like I'd just had an excellent cup of coffee, but without the caffeine jitters. I felt clear-headed and like myself, just sharper. Calmer. Without fear and without doubt. From there on, I just spent the time waiting for a problem to appear so that I could solve it.

It was only when they turned off the current that I grasped what had just happened. Relieved of the minefield of self-doubt that constitutes my personality, I was a hell of a shot. And I can't tell you how stunning it was to suddenly understand just how much of a drag that inner cacophony is on my ability to navigate life and basic tasks.

What had happened inside my skull? One theory is that the mild electrical shock depolarized the neuronal membranes in the part of the brain associated with object recognition, making the cells more excitable and responsive to inputs. Like many other neuroscientists working with tDCS, Weisend thinks this accelerates the formation of new neural pathways during the time that someone practices a skill, making it easier to get into the "zone." The method he was using on me boosted the speed with which wannabe snipers could detect a threat by a factor of 2.3.

Another possibility is that the electrodes somehow reduce activity in the prefrontal cortex—the area of the brain used in critical thought, says psychologist Mihaly Csikszentmihalyi of Claremont Graduate University in California. And critical thought, some neuroscientists believe, is muted during periods of intense Zen-like concentration. It sounds counterintuitive, but silencing self-critical thoughts might allow more automatic processes to take hold, which would in turn produce that effortless feeling of flow.

With the electrodes on, my constant self-criticism virtually disappeared, I hit every one of the targets, and there were no unpleasant side effects afterwards. The bewitching silence of the tDCS lasted, gradually diminishing over a

> period of about three days. The inevitable return of self-doubt and inattention was disheartening, to say the least. (*The Week*)

The indented paragraphs above on this page and the previous page are all from *The Week* news magazine for April 6, 2012, pages 40–41. The full article can be found at NewScientist.com.

At sixty-five years young, the silent cacophonous "minefield of self-doubt" and vexatious, almost visual noise is from all my oxymoronic rapists, who are my therapists and the main preponderating oxymoronic adjudicators, and those rapists are one and all my collective nemesis in life. Regarding "my constant self-criticism" from my August 9, 1974, "crime," the VA has been a minatorial minotaur masquerading as a menial/mentor. It is so stressful. I would like tDCS. It is just possible that the experience would be helpful and meritorious...an occasionally necessary thought out of the blue.

*Chemically balancing medications and electronic treatments even with payment augmentation are no match for the merits of my "cosmological TV." That is all that I need.*

Again, with medication, I will never be able to show if I am just as fine off my psychotropic medication as I am on the best of my psychotropic medications, even in the large secure treatment area of Napa's state hospital with its new alarm system. Lo and behold, before January 18, 2012, I was asking the right question in my request for a "psychotropic medication holiday." Truly, without the publication of my book, most of my sanity will unjustly stay an inherent contradictory oxymoronic enigma of the medicine of my censorious psychiatrist's imagination. It is currently all a constructed camouflaging catch-22 of Consta medication injections for the rest of my life. Furthermore, without being fine (on or off my psychotropic medication), it is harder to construct my "cosmic TV." So please expect me to publish my book as it is and in its entirety with emending only as is necessary.

Because of my probably superior intelligence and the information relationship between the information and news I give to and I get from my incarcerating psychiatrists, your past pejorative, prejudiced, and barely within the lower limit / standard marginal, monthly 100 percent disability payments to me (thank you) are inadequate and bad for the United States of America, just like the agnosia my low income causes in me is hard on me. I would like more money. This is an international and pivotal class action.

The bloody punitive therapy has been harming my side in court from day one. However, the bloody punitive therapy payments should not come from my side—the side that defends my United State of America under California's laws and also defends my cosmology for mankind. The bloody punitive reward should be paid to me or given to me personally because of my many attorneys-at-law and my many therapists' usual and intransigent incompetence, such that I am paid, too, for my many public psychiatrist's unusually great malice toward me. So as it seems to me, I must publish or I must perish without psychiatric pecuniary remuneration. All as I am not a medical doctor, but ipso facto my writing, I dare say that I am at that level. Furthermore, it should be paid to me at a rate that is one thousand times higher than to a regular patient because of my psychiatrist's unheard of and grievous psychiatric malice toward me.

That my unusual legal status was adjudicated on May 5, 2010, means that decision will be looked at again around May 5, 2012. I left my public defender attorney-at-law, Cheryl H. Arkansas, messages on April 5, 2012, at 9:09 a.m.; April 9, 2012, at 8:36 a.m.; April 10, 2012, at 8:36 a.m.; and other times. However, I have not heard from her recently.

Humorously noose or nous: mind, intelligence, or perception? On November 10, 1994, I wrote to then President William J. Clinton regarding the neologism below. As far as I know, he didn't act then, and now under a different democratic president, the USA has a neutered NASA.

November 10, 1994

President William J. Clinton
16 Pennsylvania Avenue
North West
Washington, DC 20500

Dear Mr. President:

Temporally in the ominous present and over the last twenty-five years, my mental health system has been maliciously unfair. Although, I had heard the word for what my mental hygiene system needs to do rather than take my constitutional and my civil rights from me and in her loving marriage to me from my perfect wife, the former Mrs. Gillian Bartholomew-Redus, until I saw the term in my newest dictionary, I did not know the meaning of the neologism "fuck off."

> "fuck off." *Webster's New World Dictionary.* New York New York: Simon & Schuster Inc., 1991. Page 544. ISBN 0-13-949298-4.
> —fuck off (slang) 2 go away!

I reverently obsecrate (on religious grounds) to your high office for help. I need my mental hygiene program to requite and FUCK OFF. Estranged and divided from my (former) wife without American liberty or American justice, I am.

Sincerely,

Mr. Dorian Gaylord Redus

My own conclusion is here the putative is just to be punitive, and our own dignity draining and destroying regressive "repressed difficulties and disagreements" afflict us all, severely worsening and aggravating Napa State Hospital's patients and staff alike.

Thank you.

Respectfully submitted,

Mr. Dorian Gaylord Redus

Mr. Dorian Gaylord Redus
Ward T-15
Napa State Hospital
2100 Napa-Vallejo Hwy.
Napa, CA 94558-6234
1(707)252-9988
1(707)255-9967

Friday, July 13, 2012

The Honorable Master Calendar Judge
San Francisco County Superior Court
Hall of Justice
850 Bryant Street
San Francisco, CA 94103

[Court Number SC088778 with Maximum Commitment Date: Until my sanity is restored]

*Regarding my treatise/book, A Quotidian Quash: From Mental Hygiene to Mental Health,* and knowing that among other things, this letter was not sent because only two of the many letters in my book were answered in writing.

*Legal Status:* I am a sixty-six-year-old light-skinned African American male under a PC 187, murder. On August 9, 1974, I stabbed a woman I had known for six years to death. October 1975, I was found not guilty by reason of insanity and remanded to Atascadero State Hospital. I was transferred to Napa State Hospital in 1982, and I was released onto outpatient status in September 1988, under the supervision of San Francisco's CONREP. After leaving, I returned to Napa from June 1994 to May 2001. Atascadero and Napa state hospitals are in California. Finally, on October 1, 2009, I returned to Napa for keeping the angry voices I had been hearing for months a secret. And on May 5, 2010, my outpatient status was revoked. Now I want to win a sanity hearing or trial this year, 2012. And after

I sent a public defender attorney-at-law my cosmological treatise, A Three-Part Discussion, on January 11, 2011, she did not return even one of the weekly telephone calls that I placed to her until Thursday, December 1, 2011. However, Tuesday and Wednesday June 19 and 20, 2012, she took me to court, and we fought an extension to keep me here at Napa State Hospital. We lost because I was the only witness for the defense. In conclusion, as I prepared for my big San Francisco Superior Court appearance on June 20, 2012, the week of Tuesday, May 8, 2012, my ward staff psychiatrist, Dr. Eric Florida, MD, said to me in his new office on Ward T-15, my 2+2=5 schizophrenic wrong answer regarding my thinking (reality) making me need my Consta medication injections every two weeks is that *I think thirty years ago, I was raped twice by abject psychotropic drugs.* Ergo, my current catch-22 situation is that my sanity off my (current) Consta injections is illusive and impossible to even desire (without my doctor and I being very uncomfortable) because he, Dr. Florida, MD, considers my twice-a-month Consta medication injections absolutely and unequivocally necessary.

Attention: Judge Garrett Maine:

I am a PC 1026 expecting a sequence of many PC 1606 or PC 1608s *in saecula saeculorum,* for ever and ever, into ages of ages, and for all eternity.

Starting in October 2010, given I (more than once) requested a "psychotropic medication holiday" at my previous psychiatrist's suggestion, and given that I am not intransigent but flexible, currently my prescribing psychiatrist, Dr. Eric Florida, MD, and I need to choose the most efficacious dose for my expensive Consta injections: 75 mg., 50 mg., 37.5 mg., 25 mg., 12.5 mg., or 0 mg. (at $566 each shot). Let's say the choosing is my phase two on Ward T-15 of this state hospital, Let's keep in mind that I have been on 37.5 mg since Friday, January 27, 2012. And let's know before and after Friday, January 27, 2012, I maintained (impeccably clean) immaculate behavior in Ward T-15's milieu under their Dr. Eric Florida, MD. Each and every day, I looked "scourly" as my first long-term psychiatrist once wrote of me in the early 1970s. Without a psychotropic "medication holiday," we will just never know. Am I in remission or in a medication-controlled and maintained remission?

For all of my life under California's aborning Department of State hospitals, what three bodies of information have been coming directly at me—one, God; two, our RCTVU (relativistic color television universe) of tomorrow; or three, the Department of State hospital's homosexual rapes? Like a corpus delicti is helpful at a murder trial, if all my lethal (behavior extracting) psychiatric experts are to stop their damnation of my fact, truth, and justice under our California Judicial System, then two out of these three bodies of information must be adjudicated as soon as possible. Videlicet, my cosmology and my having been raped are appropriate, necessary, and even mandatory for any and all my 2012 PC 1026, PC 1606, and PC 1608 hearings or trials. Furthermore, the whole truth regarding the following two issues has never come to court: (1) erudition on my private cosmologies from at least three appropriate (not appropriating) experts that all involved will readily stipulate to, and (2) the fact that three of my past abusive psychiatrists first verbally, second medically, with their brand of their abject psychiatry, did legally rape me prior to c. 1984. And third, under my past courtroom procedures, when I can "stack the whole deck" in my favor in the *un*therapeutic carnal information game between psychiatry and myself, violently or legally, I only peacefully regard the mental illnesses of my many friendly but dangerous inpatient acquaintances in this my hospital's milieu. And I (also) only peacefully regard the unreasonable and likely specific disciplinary orders and general *un*judicial guidelines that are real or imagined (powerful) carnal informational relationships that I have with my psychiatrists past, present, and future. Therefore, when I play my therapeutic "trump cards," I do not enjoy the benefit of "playing with a full deck!"

When I am obviously heterosexual, in 1981 and in 1983, because my psychiatrist's torturing psychiatric drug's horrible side effect required me to sublimate or modify my natural instinctual reaction—to fight or kill any ignoramus prescribing physician or perhaps innocent scapegoat denizen of my milieu—to the extremely deviate for me but much more acceptable practice of finding multiple (hundreds of) male sex partners on the all-male locked wards, my psychiatrist's *un*-efficacious medications, Prolixin and Haldol, were abject agents of homosexual rape! When the dangerous drugging to homosexual rape took place, it was ostensibly three things. It represented the devil. It appeared like medicine. And it was a handiwork of a minatorial human minotaur because it was rape for a year at Atascadero State Hospital by Dr. Wiggly, MD, and for a year at Napa State Hospital by Dr. Arizona, MD, 1981 and 1983 respectively! When I publish my book on it, what will the therapeutic American people's "court of public opinion" say to me about this high-stakes life and death "booty call" and (as I described above) wild card game? Will they say it was rape? I assert and proffer it is still clearly and simply up to them, a jury of my peers or a judge like you.

First, without any medication, the VA's Dr. Donald Montana, MD, with a predatory woman while I was on his VA outpatient from 1970 to 1974; second, Dr. Wiggly, MD, at Atascadero State Hospital from 1980 to 1981 with Prolixin; and third, Dr. Marshal Arizona, MD, at Napa State Hospital with Haldol. All three psychiatrists did with their dangerous power relationships verbally, medically, and through bad advice or drugs and psychiatric hospital procedures raped me prior to 1984. And if there are no such therapeutic findings, then mankind will continue to simply say I am an insane quotidian cardinal criminal in need of quashing.

Since my initial three 1966 LSD trips before my military duty (I never used any drug but alcohol in moderation in the US Army), I have been in the middle of conspicuous conspiracies since my three little 1966 LSD trips, my 1972 City College of San Francisco junior college cosmologies, and my 1974 VA homicide due to bad psychiatric advice and other academic malpractice.

As to the three (black) characters below, all being consenting, therapeutic, and African American adults (at a summer of 1966 pastime in San Francisco California), there was no sex between me and either of them, my two "guides," during that evening happening or to this day.

I said, "What is LSD like?"

He said, "It might be like a big headache. It might be like holding on to a rope for ten hours. It might be like a thin line, or it might be like a color television."

She said, "Or a party."

I said, "What happens if you let go?"

He said, "The alligators will get you. The trick is to get inside, and it might be like you did something wrong and your father is after you. If you close your eyes, then it is like the fifth dimension, but the trip is too big. Don't trip out on the bathroom!"

I said, "What do you mean?"

He said, "The colors in the bathroom, and you can get so high you pray to come down." Walking me to his small apartment's marital bedroom where there was a black-and-white TV, he said, "It might take me a while to get up enough energy, but I will be here."

From my junior college days to the present, I proffer my psychiatrists have constructed much about my life, my 1966 LSD trips, my 1972 cosmologies, and my 1974 VA homicide. In it all is their threatening, inherent, insanely paranoid view of me in their care due to the (harmful) oxymoronic information relationships between my psychiatrist's motives and my motives that make me seem schizophrenic.

## The T/reasonable Psychiatrists and the Membrane

Here it helps if you have already perused my work, *A Three-Part Discussion: (1) Recession Velocity, (2) the RCTVU, and (3) a Space-Time Sphere (4) Including Part Four* as they adduce my RCTVU theory. A Ward Q-9 staff psychiatrist called Dr. William Paul Wisconsin, MD, of Napa State Hospital was the first person to disrespect and attack my RCTVU with the hologram attack.

As I recall, it was c. 1996 in California. We were alone in his Ward Q-9 office. He was my prescribing physician, and he was wrong and eventually nullified (like he usually was) by me back then and especially here. Just like all electronic color televisions in an STS (space-time sphere) are a subset of my RCTVU (relativistic color television universe), all holograms are a subset of our RCTVU. And both are just a piece, part, or subset of our RCTVU, which is like Hermes Trismegistus's (philosophic) "the All!"

If there is a virtual "naked space-time singularity" and a virtual cosmic censorship inside our STS with a spherical two-dimensional stellar membrane conceptually just at the outer edge sandwiched in between, the central inner and the outer unknown, then the membrane may conceptually be seen as a virtual tessellation of two-dimensional quantum computer screens, hologram screens, or electronic TV screens! Three-dimensional TV is also accepted. Four-dimensional is accepted because it is time, and fifth-dimensional is also accepted as it is an STS (space-time sphere) in an RCTVU!

It is easing—necessary—to also see the importance of light (cosmic) "chi" TV and its relativistic invariance in the following *Planck* and the following empirical relativistic length and empirical relativistic time equations. In the late 1990s here at this state hospital, I used to be paid—a very few cents an hour—to screw together and unscrew big white plastic nuts and bolts. I much prefer writing this letter. It is ever more therapeutic. Again, here, I assume my readers have already seen this material when they read my *A Three-Part Discussion: (1) Recession Velocity, (2) the RCTVU, and (3) a Space-Time Sphere (4) Including Part Four.*

Our relativistic color television universe (RCTVU) must have the smaller scale "a kind of 'minimum' space-time measure (or 'quantum' of time and space, respectively), according to common ideas about [the] quantum" in an STS of an RCTVU! Although, all four of the following equations all use c or the speed or the velocity of light, and in them the $h=1.05457...\times 10^{-34}$ Joule seconds, it is a tall order to say, therefore, they are altogether based on c and evidencing QED (that which was to be demonstrated) my RCTVU theory. However, that is my comprehensible assumption or arrangement for the four separate equations as a group. Thus, the tall order of RCTVU is really to "Let there be [a tessellation of invariable] light."

The Planck time and the corresponding *Planck length* ($l_p=ct_p$) are often regarded as providing a kind of minimum space-time measure (or quantum of time and space, respectively) according to common ideas about quantum gravity:

$$t_p = \sqrt{\frac{Gh}{c^5}} \approx 5.4 \times 10^{-44} s, l_p = \sqrt{\frac{Gh}{c^3}} \approx 1.6 \times 10^{-35} m.$$

By use of these Planck units and also the *Planck mass* $m_p$ and Planck energy $E_p$ given by

$$m_p = \sqrt{\frac{hc}{G}} \approx 2.1 \times 10^{-5} g, E_p = \sqrt{\frac{hc^5}{G}} \approx 2.0 \times 10^9 J,$$

which are naturally determined (though completely impractical) units, one can express many other basic constants of nature simply as pure (dimensionless) numbers, (and these numbers may enlighten us as to RCTVU and STSs at the quantum level).

The indented above is from page 163 of *Cycles of Time: An Extraordinary New View of the Universe* by Roger Penrose, professor of mathematics at the University of Oxford.

MELT-ing Albert Einstein's equations down to prose and formula, we get:

M—mass increases with velocity. A mass in motion is proper, rest, mass divided by the radicand, one minus left hand parenthesis the velocity of motion divided by the velocity of light right hand parenthesis squared.

$$M = \frac{M_0}{\sqrt{1 - \left(\frac{V}{C}\right)^2}}$$

E—energy has an equivalent mass. Energy equals mass times the velocity of light squared.

$$E = MC^2$$

L—length in motion shortens measured in the direction of motion. Length in motion, measured in the direction of motion, is the proper, rest, length multiplied by the radicand, one minus left hand parenthesis the velocity of motion divided by the velocity of light right hand parenthesis squared.

$$L = L_0 \times \sqrt{1 - \left(\frac{V}{C}\right)^2}$$

T—time slows with velocity. The time in motion is proper, rest, time divided by the radicand, one minus left hand parenthesis the velocity of motion divided by the velocity of light right hand parenthesis squared.

$$T = \frac{T_0}{\sqrt{1 - \left(\frac{V}{C}\right)^2}}$$

Va+Vb

Velocity A plus Velocity B

Because the velocity of light is "invariable," when you add Velocity A to Velocity B, divided by one plus left hand parenthesis Velocity A multiplied by Velocity B divided by the velocity of light squared right hand parenthesis, the sum of the two velocities will never exceed the velocity of light.

$$Va+Vb = \frac{Va+Vb}{1+\left(\dfrac{Va \times Vb}{C^2}\right)}$$

With the four (micro) quantum equations, the five (macro) MELT equations, and the tessellation of the spherical membrane surrounding, covering, and encompassing all STS (space-time spheres), I adduce my RCTVU (relativistic color television universe) and not the "hologram principle."

## I Have Dodged the Following Psychiatrist's Bullets

- Some of them have said, "I will never believe Dr. Donald Montana, MD, told you to get a gun in his VA office."
- They have testified that my cosmological TV is my delusion.
- They have written that Dr. Donald Montana, MD, was not my vary "deep-pocketed" VA psychiatrist, but he was, in fact, my "private psychiatrist."
- They have written, when my only marriage was November 19, 1992, that I was re-hospitalized at Napa in 1990 because of my sudden marriage to a woman.
- Why was I in sex offender group for over three years and under investigation for over five years?
- Why did my present prescribing psychiatrist once say he is medicating me with Consta injections because I think that I was raped twice by abject psychotropic drugs in the 1980s?
- Is it that during the past forty-two years, they have been making up and constructing the falsehoods above because they have been insane and transferring their guilt and insanity to me?
- They have been making Napa State Hospital like a prison hospital of incarcerated clients like me.
- If it were not for Dr. Donald Montana, MD, at the Department of Veterans Affairs advising it, I would not

have been in the long-term relationship that led to my 1974 murder. My victim was not the only fish in the sea!
- After eighty-six one-on-one sessions with him in his VA offices, I confessed my crime to him on the day of my crime, and yet he at masterminding a very conspicuous conspiracy kept himself from coming to any of my long-term murder adjudications until he died.
- Was he both hiding his own guilt and transferring it all to me, and therefore (also) keeping the Department of Veterans Affairs free of the burden of their psychiatric pecuniary remuneration payments to me?
- Ipso facto, these are all quite queer, and they have been misguiding you and other of my San Francisco judges and misguiding good psychiatry.

---

The following pertains to me as I am almost like an elder at sixty-six years young for any elders with cerebrovascular adverse events including stroke. It may be insanely true that the "elderly patients with dementia-related psychosis treated with antipsychotic drugs [like Consta injections] are at an increased risk of death."

"*Risperdal Consta* is not approved for use in patients with dementia-related psychosis." This was from the instructions for administering *Risperdal Consta.*

When the three of us—myself, Napa staff Dr. William Paul Wisconsin, MD, and his psychiatric technician, Freda—were all notified by my then wife Mrs. Gillian Bartholomew-Redus of her decision to ask me for a divorce, I nearly or actually (once) had a stroke c. 1996. And then with Consta, I watch out for any ilk of tachycardia—or fast heartbeat—over one hundred heart beats per minutes!

When you really read my book, *A Quotidian Quash: From Mental Hygiene to Mental Health,* or your clear top page red bottom page copy of my manuscript (of the same name) which I sent to you on Friday, April 13, 2012, then you will know I really don't need my Napa's Ward T-15 staff psychiatrists making the decisions as to what dose of Consta I get.

All I need is the poetic justice for "The T/reasonable Psychiatrists and the Membrane" section, and I'd like poetic justice for the MELT equations above from my treatise, *A Three-Part Discussion: (1) Recession Velocity, (2) the RCTVU, and (3) a Space-Time Sphere (4) Including Part Four*, as they adduce my RCTVU theory.

To myself, as here and elsewhere, I use my fighting TV mind for America. I occasionally think that President Barack Hussein Obama should make me a commissioned officer—Lt. Col. Dorian Gaylord Redus, US Army, retired with pay and back pay from my 1968 discharge from the US Army. Given Tuesday, June 5, 2012, when I wrote the "The T/reasonable Psychiatrists and the Membrane" section, my after-breakfast (sitting) blood pressure and pulse rate were 125/74 and 85. And given I am not criminally insane, I am not even mentally insane, what am I doing here other than augmenting my rank?

When Dr. William Paul Wisconsin, MD, denied the therapeutic RCTVU idem above, as have others of my prescribing and diagnosing psychiatrists who probably also think that the idem above is my schizophrenic delusion causing a second (long-term rape and ignored cosmology) reasonable and unequivocal need for them to medicate me with Risperdal Consta injections. He chose his psychiatric side, making him my terrifying therapist. The (verbal) rapist who most put a black *T* in front of reasonable to spell *treasonable* in my eyes.

An STS is mathematically ergo intellectually knowable. It may be entered, and it may be virtually escaped. Therefore, an STS is a virtual *naked* space-time singularity. What is inside is "here." The inside and the outside are one and panspermic under the ubiquitous laws of our astrophysics. The evolution inside is academic and knowable. All of STS's information is (in principal/theory) in the membrane. And furthermore, the membrane is like a tessellation of plastering spherical 2D holograms, tiles, or more correctly, color television screens, in general, Planck-sized two-dimensional RCTVU units—no matter what psychiatrist Dr. William Paul Wisconsin, MD, once said to me in private in his office in 1996!

Below is a quote from page 44 of the course guidebook of UC Berkeley's Dr. Alex Filippenko's *Black Holes Explained*, course no. 1841, available from *www.thegreatcourses.com*—DVD lecture no. 11.

One way to view this conclusion is through the holographic principle: All of the information inside a black hole is actually contained on its surface, a thin membrane just outside the event horizon that resembles a hologram. This is the membrane paradigm introduced in Lecture 8. Recall that from the perspective of an outside observer, nothing ever crosses the event horizon of a black hole, due to infinite time dilation. The material accumulates on a very thin membrane just outside the event horizon; this is sometimes called a stretched horizon. Since all of the information in a black hole can be thought of as being on the thin, 2-dimensional membrane on the event horizon, this is kind of like a hologram. A hologram is a 2-dimensional plate of glass, or other [2 or 3-dimensional color television screen] recording device, which when illuminated, shows a realistic 3-dimensional image. By shinning light on a sequence of holograms, one can even make a 3-dimensional movie.

In this case, the membrane has a thickness of only 1 Planck length (about $10^{-33}$ cm), 20 orders of magnitude smaller than a proton! According to the membrane paradigm, the information was never inside the black hole; thus, it can in principle escape. To an outside observer, the membrane is very hot and emits particles that somehow carry the original information. Quantum mechanics is saved! However, the information has been scrambled; it has high entropy.

On page 45, Professor Alex Filippenko goes on to say, "Mathematically the contents of the entire visible Universe might be described by information plastered in Planck areas at its visible edge!"

If one sees that "no phenomenon is a phenomenon until it is an observed phenomenon" (John A. Wheeler), and if one asks, "What is an observer? [thinking] Actually any physical record will do" (Dr. Benjamin Schumacher), then even a prescribing "ignoramus" may see "If I am the only factual record keeper of dangerous medical and forensic malpractice, then knowing the malpractice is unrecorded without me, and then knowing, said, malpractice instances, even many of them, have no policing public conscience without my book, one should decide to buy my book, and decide for my book's information to be recorded by my many probable readers reading it."

The (indented) blurb below is for the back of my first book's hardbound cover jacket and back cover of its paperback version.

> This manuscript was not supposed to go public. However, because it was written in three joined parts and over two consecutive years, there was a problem as to what to call it chronically, 1969–2011 or 1969–2012. So the author chose the latter.
>
> Author, Mr. Redus, has looked into the cathode ray tube of his soul to find his way out of his "commonplace silencing." (It is what *A Quotidian Quash* really means). Actually, in his book, the evolving cosmological TV energy of his puzzling psychiatric history from junior college years to decades as a psychiatric inpatient unfolds and jumps. *A Quotidian Quash* also puts forth two new cosmic theories, RCTVU (relativistic color television universe) and STS (space-time sphere), as Mr. Redus also "blows the whistle" on California's Department of State hospitals for being a psychiatric minatorial minotaur masquerading as a menial mentor.
>
> There is evidence of his many past psychiatrists' paranoia in this book, but when his psychiatric future arrives, his tomorrow, comes to us as his RCTVU theory, pristine and virgin, it is excellent and okay as it puts itself in his creative hands. Moreover, it requires that we have learned

something from our evolutionary yesterday or from God in and subtly outside of his book's other speculative theory, STS theory.

This is a book of many therapeutic letters mostly to his California therapists. And in it, we learn that they were so very interested in recording "something wrong" with him, their long-term psychiatric patient, Mr. Redus, that there became "something very obviously paranoid and wrong" with how all of his California therapists treated him, their long-term psychiatric patient, from the start of his treatment to the finish of my treatment with them.

This email that will go to hundreds of thousands of prospective buyers is the only completely finished part of my book project:

*A Quotidian Quash: From Mental Hygiene to Mental Health*
By Dorian Gaylord Redus

Dorian Gaylord Redus has looked into the cathode ray tube of his soul to find his way out of his "commonplace silencing." (It is what *A Quotidian Quash* actually means.) In this book, the evolving cosmological TV energy of his puzzling psychiatric history from junior college years to decades as a psychiatric inpatient unfolds. *A Quotidian Quash* puts forth two new cosmic theories, RCTVU (relativistic color television universe) and STS (space-time sphere), as he, Gaylord, also "blows the whistle" on California's aborning Department of State hospitals for being a minotaur masquerading as a mentor.

In past courtroom testimony, past nontherapeutic consults, and opinions, my psychiatric adversaries convincing denial is a vexatious flabbergast that pins psychiatry's insane donkey tail on me, its sane long-term patient. This is a very wicked (prank-like) "transference" of psychiatric guilt to me by, for example, choosing one of the following for me: 0 mg., 12.5 mg., 25 mg., or 37.5 mg. (where I have been since Friday, January 27, 2012), or up even higher to 50 mg. or 75 mg. of injectable Consta medication injections every other week, as if I am in a medication-controlled remission.

The Department of State hospitals virtually says my insanity anywhere is my insanity everywhere. I say, their virtual quotidian quash of my sanity anywhere is a squashing of my sanity everywhere. Don't let my psychiatric adversaries give rape a good name! As I have outlined above, I am demonstratively sane, and the demonstrativeness of my sanity is not a problem!

Finally, against the Dr. Wisconsin effect (attack) with the "holographic principal," Alex Filippenko says on page 45 of his course guidebook for the DVD series *Black Holes Explained*, "So there is almost certainly no physical hologram at the [or at our STS's event] horizon, like some giant machine clanking away." The only one Napa's Dr. William Paul Wisconsin, MD, was helping was the devil!

---

Please decide for my RCTVU the immediate restoration of my legal sanity and for my complete intellectual and financial freedom.
Thank you.

Respectfully submitted,

Mr. Dorian Gaylord Redus
Psychiatric patient

Mr. Dorian Gaylord Redus
Ward T-15
Napa State Hospital
2100 Napa-Vallejo Hwy.
Napa, CA 94558-6234
1(707)252-9988
1(707)255-9967

                                          Thursday, August 9, 2012

Dr. Shelly Massachusetts, PhD
Anka Behavioral Health Services
Golden Gate Conditional Release Program
350 Brannan Street, Suite 200
San Francisco, CA 94107
1(415) 222-6930

Re: Your 5-4-12 Hospital Liaison Report due to our Wednesday, April 11, 2012, meeting, among other things, like permission to publish CONREP's April 12, 2010, letter to revoke me on May 5, 2010.

*Legal Status:* I am a sixty-six-year-old light-skinned African American male under a PC 187, murder. On August 9, 1974, I stabbed a woman I had known for six years to death. October 1975, I was found not guilty by reason of insanity and remanded to Atascadero State Hospital. I was transferred to Napa State Hospital in 1982, and I was released onto outpatient status in September 1988, under the supervision of San Francisco's CONREP. After leaving, I returned to Napa from June 1994 to May 2001. Atascadero and Napa state hospitals are in California. Finally, on October 1, 2009, I returned to Napa for keeping the angry voices I had been hearing for months a secret. And on May 5, 2010, my outpatient status was revoked. Now I want to win a sanity hearing or trial this year, 2012. And after I sent a public defender attorney-at-law my cosmological treatise, *A Three-Part Discussion*, on January 11, 2011, she did not return even one of the weekly telephone calls that I placed to her until Thursday,

December 1, 2011. However, Tuesday and Wednesday, June 19 and 20, 2012, she took me to court, and we fought an extension to keep me here at Napa State Hospital. We lost because I was the only witness for the defense. In conclusion, as I prepared for my big San Francisco Superior Court appearance on June 20, 2012, the week of Tuesday, May 8, 2012, my ward staff psychiatrist, Dr. Eric Florida, MD, said to me in his new office on Ward T-15, my 2+2=5 schizophrenic wrong answer regarding my thinking (reality), making me need my Consta medication injections every two weeks, is that *I think thirty years ago, I was raped twice by abject psychotropic drugs.* Ergo, my current catch-22 situation is that my sanity off my (current) Consta injections is illusive and impossible to even desire (without my doctor and I being very uncomfortable) because Dr. Florida, MD, considers my twice-a-month Consta medication injections absolutely and unequivocally necessary.

To Whom It May Concern:

Are we aborning? I am a PC 1026 expecting a sequence of many PC 1606 or PC 1608s *in saecula saeculorum* for ever and ever, into ages of ages, and for all eternity, unless CONREP (inter alios) halts and becomes part of the solution. Is CONREP amenable? CONREP's compulsion to resist my corrections or instructions and call them my anger is CONREP's not being amenable to my truth and my fact. And moreover, when right and wrong are unknowable in court because they (court officers) "lie double entendre" above the law in court with complete CONREP discretion, they pin the Department of State hospital's criminally insane donkey tale on me and my side *un*justly!

The (therapeutic) excellent fact is that your May 4, 2012, CONREP Hospital Liaison Report due to our Wednesday, April 11, 2012, meeting quotes me (on my rape) at the top of its page one in its first paragraph. When to be clean and not dirty, it is true that I was homosexually raped in the 1980s. Your page 1 quote is largely a super great beginning. However, you, CONREP, also mistakenly make the nasty and arrogant assertion that I continue to be "paranoid" and "delusional" in thinking "certain psychiatrists and [their] medications might have caused [my]…homosexual acts."

In my paragraph, your use of the word *abhorrent* where I used the word *aborning* to describe my "aborning rage" in the 1980s, on my state hospital psychiatrist's torturing medications at Atascadero State Hospital and at Napa State Hospital, is your Sigmund Freudian slip that adverts to your correct and real therapeutic feeling about my de facto rape. Moreover, when you two mistakenly write to transfer it, your denial, delusions, and paranoia to me in a bizarre and an assertive attack or nontherapeutic written *coup de grace* as if I am the brunt or target of your argument, I see your de facto, denial, delusions, and paranoia about your slip and my homosexual rape. (See my indented paragraphs below.) My psychiatrist's actual anger is the source and cause of the raping. I hope in your use of my paragraph that your erratum may be used to advert. I hope you seek the truth and the fact, which is that the *abhorrent subject* of my quoted paragraph—from my proposed book on rape, inter alia—is indeed an abhorrent and important issue to us. Until your report, all the-rapists, my therapists—all at a quotidian quash of all my information on their rapes from the 1980s to 2012 at the hospitals I depend on most—have, except for your quote of my paragraph, written nothing (to me about the rapes) that comes to my mind at this writing. Before you two, there was a complete and utter quotidian quash making me appear almost schizophrenic on the abhorrent and ineluctable subject of my VA heterosexual and Department of State hospital homosexual rapes by my mostly psychiatric therapists, the rapists, before 1985! Editing this paragraph on 7-28-12, my sitting 7:45 a.m. blood pressure and pulse were 113/74 and 80, respectively, much better than my usual after-breakfast numbers. But when I pray that you see the characters all have camouflaging, so the rapists, my psychiatric therapists, don't look like the psychiatric predators they are, and I, their long-term victim, won't look like their prey. I am paranoid that the camouflaging is called San Francisco's County Superior Court.

Arguably, in your May 4, 2012, CONREP Hospital Liaison Report, your mention of my long-denied homosexual rape in the opening paragraph evolves me in a key "information relationship" that has unnecessarily, inappropriately, and recently been a stated way for my prescribing psychiatrist, Dr. Eric J. Florida, MD, to mandate my being medicated with 37.5 mg. of Consta (Risperidone injectable) every other week from 1-27-12 ongoing *in saecula saeculorum*, for ever and ever, into ages of ages, and for all eternity. When Dr. Eric J. Florida, MD, lost it, it is (also) true twice at a brinkmanship to stop me, he argumentatively promised me something from him in writing on my homosexual rape during a treatment team and in a formal discussion in his office. He gave me nothing but trouble as he lied to me both times!

If I correct CONREP and Napa State Hospital, I am *not angry* here. It may be a vexation to CONREP and Napa State Hospital. It may cause you/them to manipulate, diagnose, and write of my supposed or putative anger, but that is just your/their pathological casuistry. The anger here is dichotomous and not mine, but it is anger. You're not being inappropriate authorities with a complicity in the Department of State hospital's (harmful) oxymoronic, incongruous, and contradictory therapy by covering up my rape with denial makes you two authors—of your May 4, 2012 Hospital Liaison Report—a pair of heroic heroines that feel/look to me like the older nineteen-year-old Alice in Wonderland in her silver armor, with her (inclusive) vorpal sword on Frabjous Day, slaying a heinous Jabberwocky of exacerbating intellectual denial of rape and other things like my cosmological TV theory! I see your elan in my new 3D Disney DVD of *Alice in Wonderland* starring Johnny Depp as the Mad Hatter.

When I am obviously heterosexual, in 1981 and in 1983, because my psychiatrist's torturing psychiatric drug's horrible side effect required me to *sublimate* or modify my natural instinctual reaction—to fight or kill any ignoramus prescribing physician or perhaps innocent scapegoat denizen of my milieu—to the extremely deviate for me but much more acceptable practice of finding multiple (hundreds of) male sex partners on the all-male locked wards, my psychiatrist's *un*-efficacious medications Prolixin and Haldol were abject agents of homosexual rape! When the dangerous drugging too homosexual rape took place, it was ostensibly three things. It represented the devil. It appeared like medicine. And it was a handiwork of a minatorial human minotaur because it was rape for a year at Atascadero State Hospital by Dr. Wiggly, MD, and for a year at Napa State Hospital by Dr. Arizona, MD, 1981 and 1983 respectively! When I publish my book on it, what will the therapeutic American people's "court of public opinion" say to me about this high-stakes life and death "booty call," and (as I described above) wild card game? Will they say it was rape? I assert and proffer it is still clearly and simply up to them, a jury of my peers or a judge like you. (From my Friday, July 13, 2012, unsent letter to the Honorable Garrett Maine Judge Superior Court San Francisco)

## Isn't It Rape?

Isn't it homosexual rape by the numbers? Isn't it just one midteens homosexual act with a Black cousin and just one sudden early twenties homosexual act with an older White man (while I was both getting out of the US Army and living at home with my parents) in San Francisco California? Isn't it one hundred homosexual mates in one isolated year on Prolixin at Atascadero State Hospital all on their Ward 27 and one hundred homosexual mates in one isolated year on Haldol at Napa State Hospital on their (combined) Ward Q3 and Q4 all before 1985—to say nothing about my post US Army and my outpatient VA heterosexual sex before 1975? Isn't it all just rape by the numbers?

When they were pathological, inappropriately psychiatric authorities may have thought I was ruminating on evil sin, but the only sin at issue was in my psychiatrist's worrisome thought that I was pathological, not them. Here at this state hospital, ensconced on its Ward T-15, I am free to showcase the "information relationships" between my psychiatrist's motives and my patient motives, like in the indented paragraphs above, so I am still no longer hearing the angry (alliterative) vicious voices that caused my return to Napa on October 1, 2009.

Re: The grandiosity in your "Mr. Redus has written numerous letters to CONREP since his rehospitalization, content being grandiose, angry and paranoid." I have an acute credibility problem. However, I will not put up with your inappropriate appropriating put-downs.

## My Grandiosity

- In 1972 to 1974, American science lost sight of its L5 space colonies.
- My letter to Dr. Martin Luther King Jr. was an unsent attention-getter full of "reverse neglect" of the fact that I was raped.
- I took my last LSD with pathos, regrettably murdered a woman I had known for six years, and Donald Johanson found "Lucy," an *Australopithecus aferensis* in 1974.
- My great marriage began to fail, and Tim White, a paleo-anthropologist at the University of California, Berkeley, found Ardi, an *Ardipithecus ramidus*, in 1994.
- I wrote the "T/reasonable Psychiatrist and the Membrane" section of my Friday, July 13, 2012, letter to the Honorable Judge Garrett Maine, San Francisco Superior Court. Albeit, it was (also) unsent.

They are not helping me, and they are exacerbating for me, but my contention is that I am not in remission to sanity because of my low-profile outpatient treatment programs. I am not okay because of my hospital's low-profile therapeutic groups, and I am not okay because of my last low-profile San Francisco Superior Court appearance. I also contend that the advent of my high-profile "whistleblowing" book on California's Department of State hospitals and the VA, which I wrote to stop their de facto raping of my body that started thirty and forty years ago, that book supports my remission to sanity. And furthermore, my book/work advocates that my remission may not be medication dependent, controlled, and maintained as previously and putatively supposed.

## The Onus of the Word Kill

I must feel the evolving nontherapeutic "information relationships" that are put in and out of commission in the following almost humorous anecdotal incidents.

The onus is my psychiatrist's problem and not my problem, even though psychiatry has transferred its exacerbating onus to me into the ages of ages through malpractice too oblivion at hospitals and in San Francisco's County Superior Court from 1969 to 2012.

For me to have my individual RCTVU (relativistic color television universe) from its inception to its theoretical proof completely ignored is like being blown away by an astrophysicist's big bang as one blows out a candle's light. For me to have my psychedelic individual's LSD trips from their initial ingestion completely ignored is like being blown away by an nontherapeutic Veterans Administration psychiatric outpatient heterosexual raping, and it is like being blown away by an nontherapeutic California Department of State hospital's psychiatric inpatient homosexual raping like one blows out a religious candle's light and hope. Consequently, it is very important that I stop and make sure it is most important that we are all human beings working with human beings, and that we are never an individual humanoid hiding behind humankind, as an assuaging group, and doctoring "human things." A "human thing" is a client who's psychiatrist is hypocritically applying his doctor's Hippocratic oath, and seeing the paying client's face as a Hippocratic face, sequestered in his doctoral oblivion, checked only by his oblivion and balanced only by his or her oblivion.

1. I am intentionally underpaid and soundly put down by psychiatry.
2. Ms. Edna Ella Robenson, whom I killed, was criminal, and Dr. Donald Montana, MD, was my crazy VA psychiatrist.
3. In reality, before I murdered, I was at a very necessary brinkmanship in the VA clinical report below. I needed my lover, Ms. Edna Ella Robenson, to politely acquiesce, choose the very romantic safety of our sexual relationship first, and second, stop her regular unilateral violence against me, conceding a peace between us out of her love in our relationship. Therefore I, like a domesticated policing person in the VA clinical report below, gently showing (but not threatening at all to use) my "deadly force," tried to dissuade and to stop Edna's far too frequent and sexually crazed violence. Again, I just needed to stop Edna from regularly pulling our kitchen knives on me out of the blue. So I was very much on the side of keeping domestic peace. I was at my wit's end because of her knife fighting and, at the limit of my usually inexhaustible mental resources, utterly at a loss. (I had reported everything to my Veterans Administration psychiatrist, Dr. Donald Montana, MD, but he just advised me to "marry her.") So I reiterated and recounted everything to a second VA therapist at a VA hospital, and the floridly psychotic second VA therapists, unable to tell "right from wrong," wrote the following and exacerbating VA clinical report:

> Most recently patient [Mr. Dorian Gaylord Redus] has been becoming increasingly angry and hostile. He stated that he pulled a spear off the wall, broke it, and had her [Ms. Edna Ella Robenson] backed up against the wall with the half of the spear across her throat, pinning her to the wall. When he realized that

he could easily kill her, he became frightened of his own behavior and decided to return to the hospital. Patient has a comfortable income. He is 100% service connected and receives $495 V.A. Compensation. (From a 2/2/73 clinical record by Ward Psychologist Lois A. Connecticut, PhD, and Ward Physician Peter L. New Hampshire, MD, page 2)

4. In their April 12, 2010, letter to my court, Dr. May, PhD, and Christopher A. Idaho, LCSW, said requesting my outpatient treatment be revoked:

> He [Dorian Redus] had informed his psychiatrist at the time that his girlfriend was having an affair. He heard his psychiatrist respond, 'Get a gun and shoot her,'" (from page 1).

And they also said on page 4:

> On several occasions, he thought he heard his psychiatrist [Dr. Donald Montana] telling him he should buy a gun to kill her [Ms. Edna Ella Robenson] and that his psychiatrist was coaching him to commit murder.

These two small quotes are not only more grievous and serious than we suppose, but they are more grievous and serious than we can suppose. Furthermore, when I was visiting her just before I took her life on the worst morning of my life, I found Edna on her telephone, in her sheer off-white bra and panties. And after Edna got off the telephone, she said to me, "That was the police. They are getting me a gun." Then Edna said she was going to kill me with the gun—expletive deleted. *And what I said, so help me God, about the VA's Dr. Donald Montana is that once or twice in his VA office, he told me, "Get a gun."*

5. On 27 April 1973 I saw Mr. Dorian Redus in psychiatric consultation at his request... He [Mr. Dorian Gaylord Redus] asserted that Dr. Donald Montana, his [VA] therapist, had told him to get a gun in case he needed to use it against his girlfriend's other boyfriend. The patient is quite delusional on this point. Knowing [the VA's] Doctor Montana I know that no such advice was given and in talking briefly with Doctor Montana he told me that he was aware of the patient's delusion about alleged advice for the patient to get a gun. From a May 3, 1973, letter to the Veterans Administration by psychiatrist Dr. David W. Delaware, MD, page 1)

6. In their April 12, 2010, letter to my court, on page 4, Dr. May, PhD, and Christopher A. Idaho, LCSW, said requesting my outpatient treatment be revoked:

> He compulsively organizes his medications to check and double check his compliance by writing down the date and times he takes his medication. At times when he feels heightened pressure, his anxiety intensifies his thought disorder and he appears less organized in his speech and becomes tangential. His anxiety may then turn into distrust and apprehension. If his thinking goes unchecked with the support of his clinical team, he may become fearful and mildly paranoid. *An example of this includes a description made by Mr. Redus before the treatment team (on September 28, 2009) when he was confronted about his intrusive thoughts. He stated that when he is alone, there are times when he has thoughts of preventing self-harm by taking precautions. He has two windows in his bedroom, where one has a screen and the other does not. He [Dorian Gaylord Redus] makes sure that he does not change his clothes in front of the window without a screen for fear that he may fall out accidentally.*

I vehemently disagree. It is my observation that my thought disorders are exacerbated by my CONREP team. My clinical team, i.e., Dr. May, is a cause of some of my fear. See the onus of what Dr. Wendy May, PhD, did not choose to say, videlicet:

- First, regarding the windows, when I made my private comment to her about my not dressing in my window (which still seems to me to be a good decision), Dr. Wendy May and I were alone in my private bedroom in my, like her, Asian neighborhood.
- Second, CONREP client Ms. F, my best lady friend at CONREP San Francisco, had just days or weeks before jumped from a window of her home and nearly died. And moreover, she had been coming over to socialize with me, very religiously, almost every Sunday, for months. We would have dinner and watch a movie every Sunday at my bachelor apartment. But one night, she very mysteriously decompensated and relapsed, perhaps from personal family problems. I was daunted, devastated, and I somewhat felt let down.
- Third, CONREP client Mr. S, my housemate with whom I shared a CONREP bedroom from 2001 to 2003, had also just months before fallen or jumped from his (but not one of my) bedroom windows somewhere in San Francisco, California.

7. In their April 12, 2010, letter to my court, on page 3, Dr. May, PhD, and Christopher A. Idaho, LCSW, said requesting my outpatient treatment be revoked:

> A CONREP clinician went to visit him [Dorian Gaylord Redus] at Napa State Hospital on December 16, 2009 to assess his mental status to determine whether he is appropriate to return to the community. After a lengthy visit, Mr. Redus

shared that he had thoughts about giving his CONREP clinician (whom is currently pregnant) a Cesarean-section and stated that he had these thoughts since clinician's first pregnancy in 2008. When explored further, Mr. Redus said he did not have any intentions of harming his clinician [Dr. Wendy May, PhD].

- Dr. Wendy May, PhD, was a popular clinician around the CONREP office. She was also in my apartment once a month to make a home visit and check my strictly controlled internet use.
- Albeit, the figment of my imagination or inkling about giving a doctor a Cesarean section only happened once. Adam, one of Napa State Hospital's social workers, was so programed as to my evil pathology and guilt by CONREP, he programed me to confess of the almost humorous and not evil anecdotal Cesarean section incident on December 16, 2009.
- Dr. Wendy May, PhD was an obviously pregnant bride and not a very good role model!
- When my ex-wife and I first married suddenly, she was not a pregnant bride. Although, we did get hitched up, married, suddenly. From our first date to our Hall of Justice wedding, it was only thirty-three days, as I recall our November 19, 1992, civil ceremony.

8. Although, the recording therapist here omits my lifelong relationship with my only daughter's mother that started in high school and ended with Marilyn Robinson's death. She recorded the following:

> Mr. Redus has been involved in three long-term relationships with females. The first relationship was with his eventual victim. The second relationship occurred when Mr. Redus was first

released to Conrep. He met her in school [City College of San Francisco] and married her 33 days after their first date. All done without the knowledge or approval of Conrep, which eventually led to his rehospitalization. The third relationship occurred during the summer of 2003, when Mr. Redus became interested in a female whom he met and befriended at school. He continued his friendship with the woman [Asian student], and during one of their meetings, he gave her a back rub, kissed her, and touched her bare breasts. After repeated encouragement from staff, Mr. Redus told this woman about his crime and apparently had not heard from her since. (From page 1 of a 14-Day, Ward T-2, WRP dated 11-24-2009)

- After I told her of my 1974 homicide, our friendship continued. She stopped seeing me (I believe) when CONREP insisted that I tell her about my sex offender treatments, which were a de facto embarrassing and unreasonable waste of my time as they were private. And furthermore, there was no initial offense, recidivism of offenses, or recrudescence at all.
- Robin H. was my Japanese student/friend for a year. During the romantic encounter, she intimated to me that she had delivered her (then grown-up) son through a Cesarean section. Therefore, my split-second inkling in my Chinese Dr. May's office, as I saw a yellow pencil, is just not as pathological as CONREP sees it and would have everyone believe. Both were nubile students when I met them.

9. I have had deeply seated repressed regression due to my unvoiced disagreements regarding the following stuff by Dr. Louisiana, PhD, who formerly consulted out of his CONREP office in San Francisco. As he is quoted by my

CONREP clinician, Dr. Wendy May, PhD, I vehemently and intellectually disagree with their conclusions about me and my relationship with my woman. So I ask, who is actually guilty of sin—my therapists, the rapists, or me, their model patient? How could it be me or my ex-wife? When being lied about, I do not deal in lies.

> Whenever Mr. Redus begins to become involved with a female, it is important that this relationship is monitored closely. As stated in the most recent psychological evaluation, if there is any one area wherein Mr. Redus may unravel and cease to take his medications, it is in the area of interpersonal relationships. He rather naively tends to downplay the extreme friction he encounters with women and instead emphasizes his warm, loving, caring nature. Nevertheless, Mr. Redus needs to continue to work on how to participate appropriately in a relationship by maintaining stable boundaries and avoiding sexual inappropriateness. [paragraph] According to the most recent psychological assessment (dated 5/2006), results indicate that Mr. Redus was "bent on presenting himself in the most favorable light possible as he essentially denied all problems, both large and small." Per Dr. Louisiana, evaluator, "data suggest that Mr. Redus' more egregious psychiatric symptoms are largely in a state of medication-controlled remission." However, "Mr. Redus has unusually strong and developmentally regressed dependency needs. Should such needs be energized by the strong emotionality beneath his surface; it is questionable whether or not Mr. Redus would be able to control himself." Although, he was cooperative during the assessment, "one worries both about his stability and his unconserved anger should he

ever get into a serious intimate relationship which begins to turn problematic." Mr. Redus has a severe mental illness. The symptoms of which—delusion, paranoia, and anger—are currently maintained by medication. If Mr. Redus engages in interpersonal relationships, particularly with women, Mr. Redus may harbor the most potential for provoking emotional conflict resulting in stopping his medication and decompensation.

The previous quote is from an April 12, 2010, letter to my former court by Dr. May, PhD, and Christopher A. Idaho, LCSW, and it can be found on page 5. Dr. Louisiana's report/work is like CONREP, a "sacred cow." The two are both immune from criticism. However, again, from page 4 of this 8-9-12 letter, for my intrinsic reality-based remission too sanity, my book/work advocates that my remission may not be medication dependent, controlled, and maintained as previously and putatively supposed. I do, therefore, request (again) an appropriate psychotropic "medication holiday." When I fend, defend myself, and I substitute a secret (but plausible feign of fiendish and feisty) drug use on my former friend, Ms. Edna Ella Robenson, by Dr. Montana (inter alios) for Dr. Louisiana's mandated "medication" for me in CONREP's letter, then Dr. Louisiana (there) may well be evaluating my former long-term VA psychiatrist, Dr. Donald Montana, MD, with, in, and through me. Yes! Dr. Louisiana, PhD, may have actually made a de facto diagnosis of things that I did because of Dr. Montana's private behest and later public denial to save his White skin and avoid his Black onus. Videlicet, I know he told me to get a gun! This is all deeply embedded in my massive malpractice argument. Admittedly, it is there argumentatively. It took me thirty-eight years to get a clear document out of the blue from the Veterans Administration that documents that I gave Dr. Montana LSD in one of his VA offices before the untimely demise of my victim, Ms. Edna Ella Robenson, and my giving her and her sister Ms. R LSD. Primarily due to my undignified relationship to my nullifying and adversarial therapists, the rapists, I am not presenting all my reasonable excuses regarding the April 12, 2010, CONREP letter. My hospital and my CONREP have all my prosecutors safely in their pocket.

10. From mental hygiene to mental health, the highest adjudicator is the defendant over the rapists. The onus is not only on the rapists, it is on my defense counsels, my honorable judges, and last but not least, on my prosecuting counsels. And therefore, my book is at my behest also going to the American people as a threat I hope to make a promise.

I am poised here on Ward T-15 to make a pristine or virgin academic quantum leap up to appropriate communication and public forensic policy regarding my private therapy. I am at publishing and selling my book, *A Quotidian Quash*, but I need your essential written permission for my publisher, Trafford Publishing, to publish, as part of my proposed book, one single letter you wrote about me.

Please (also) help me again by sending me CONREP's letter of permission, approval, and authoritative approbation for my publisher to publish CONREP's April 12, 2010, CONREP report to court and revocation of our outpatient treatment relationship letter to the Honorable Judge Wyoming of San Francisco's County Superior Court. I sent you the (six-by-nine-inch inset photocopied) pages from my proposed book, *A Quotidian Quash: From Mental Hygiene to Mental Health 1969–2012*.

Nick Seattle
Production Finalization
Trafford Publishing
Suite 200
1663 Liberty Drive
Bloomington, IN 47403
1(888)232-4444 Ex 6124
Web: *http://www.trafford.com*

Please address your letter to Nick Seattle as above, and send it to me so I may send it to him, as he has requested your permission for him to publish in a letter.

Thank you for your assistance.

In conclusion, it has been four months since our last meeting, and in my imaginary Super Bowl, you, CONREP, continue to get my MVP for running the wrong way with my therapeutic TV football. And you therefore, make my therapeutic TV football *un*therapeutic. If you come and see me here at Napa State Hospital on Ward T-15, then we may discuss this letter. In further concluding arguments, before I began publishing my book (to me), the onus or burden and blame for this my fiasco was on the late great Ms. Edna Ella Robenson, as privately evolved and empowered by the Veterans Administration's psychiatry. However, the punishing public *onus pro bandi* or obligation in the future is transferred to the American people, where it ultimately belongs, as I am publishing soon.

Thanks again.

Respectfully submitted,

Mr. Dorian Gaylord Redus

# A QUOTIDIAN QUASH'S EPILOGUE

There are three parts to *A Quotidian Quash*. Part One ended with an appealed denied; a list of dates of forty-eight conjugal visits, all officially nonsexual; a poem about a laughing lady; and a reading test with answers. Part Two ended with the perfidious and short story about the hospital trust officer and my stereo, but we can't forget my million-dollar debt to Napa State Hospital. And at writing the end of Part Three on President's Day, Monday, February 20, 2012, early in the morning at 4:30 a.m., while I was waiting for the only working shower on Ward T-15, the day after I had half arranged to be published, I wrote that I dreamed that I was in the United States Army. I was stationed in San Francisco California. And I also wrote that in the dream, I was issued a color TV as I was about to phone family in the city. In the dream, I also attacked a nemesis who had stolen my Army-issue TV, and in that attack, I dislocated the jaw of my nemeses with the heels of both of my hands. Finally, I actually awakened from my dream as the face of my nemesis stretched became leathery in the dream, and he just disappeared.

This is, of course (if and only), if my medication is absolutely necessary. However, the cornerstone and best part of *A Quotidian Quash*, which means "a commonplace silencing," is if it is raining outside of my abode—and my psychiatrists are using an adversarial tough love in their abject psychiatry to make me "wise up" and get on their queer bandwagon of homosexuality instead of thinking for myself, and especially if my medication regimen is *un*tolerated by me before my RCTVU and my STS theory beliefs are treated by said *un*tolerated medications—we should see if it is actually raining outside, or at least not treat me as if I am a delusional nutcase without an appropriate and consulting cosmologist's and astrophysicist's testimony on my brilliant RCTVU (relativistic color television universe) and STS (space-time sphere) theories.

Thank you very much.

Society wrecked my sex life, made my "cosmic TV" illegal, and they transferred all their guilt to me.

# ABOUT THE AUTHOR

Dorian Gaylord Redus at age 33

Mr. Dorian Gaylord Redus currently resides in San Francisco, California, with his daughter, two grown granddaughters, and his new grandson-in-law.

He was in two California state hospitals, and he was on one San Francisco community outpatient treatment program for all his forty-five annual, consecutive, and abusive stellar trips around our star, the Sun. All the way, he was putatively delusional, according to his psychiatric caregivers concerning his junior college cosmology and astrophysics. Moreover, after he was freed by his daughter in 2019, in 2020, he went to live with her. His greatest problem has been (and is eternally) to accept that it is a "piece of chocolate cake" to stay out. However, his psychiatry—that he was under (1975 to 2020) at the two state hospitals and on the one outpatient treatment program—never let on that his getting out and staying out was always "easy as pie," for all concerned. Therefore, he was a big-time long-term prisoner of his psychiatry's requirements. Financially, just before he left his outpatient program, left all his controlling forensic holds, and just before he let go of his hospital's psychiatric caregivers, "The average annual cost of one forensic patient nationwide was $341,614" on October 1, 2017.

Lightning Source UK Ltd.
Milton Keynes UK
UKHW010713191222
414157UK00001B/23